T0314499

Intumescent Coating and Fire Protection of Steel Structures

Intumescent Coating and Fire Protection of Steel Structures establishes the thermal insulation characteristics of intumescent coating under various fire and hydrothermal ageing circumstances and shows how to predict the temperature elevation of steel structures protected with intumescent coatings in fires for avoiding structural damage.

Introduced are the features and applications of intumescent coatings for protecting steel structures against fire. The constant effective thermal conductivity is defined and employed to simplify the quantification of the thermal resistance of intumescent coatings. An experimental investigation into the hydrothermal ageing effects on the insulative properties of intumescent coatings is presented, as well as the influence of a topcoat on the insulation and ageing of such coatings. Also described is a practical method for calculating the temperature of the protected steel structures with intumescent coatings in order to evaluate the fire safety of a structure.

The book is aimed at fire and structural engineers, as well as researchers and students concerned with the protection of steel structures.

Intumescent Coating and Fire Protection of Steel Structures

Guo-Qiang Li
Ling-Ling Wang
Qing Xu
and
Jun-Wei Ge

CRC Press is an imprint of the
Taylor & Francis Group, an **informa** business

Cover image: Jun-Wei Ge

First edition published 2023
by CRC Press
6000 Broken Sound Parkway NW, Suite 300, Boca Raton, FL 33487-2742

and by CRC Press
4 Park Square, Milton Park, Abingdon, Oxon, OX14 4RN

CRC Press is an imprint of Taylor & Francis Group, LLC

Library of Congress Cataloging-in-Publication Data
Names: Li, G. Q., author.
Title: Intumescent coating and fire protection of steel structures / Guo-Qiang Li, Ling-Ling Wang, Qing Xu, Jun-Wei Ge.
Description: First edition. | Boca Raton : CRC Press, 2023. | Includes bibliographical references and index.
Identifiers: LCCN 2022047189 | ISBN 9781032263564 (hbk) | ISBN 9781032263571 (pbk) | ISBN 9781003287919 (ebk)
Subjects: LCSH: Building, Fireproof. | Building, Iron and steel. | Steel--Thermal properties. | Insulation (Heat) | Thermal barrier coatings. | Thermoresponsive polymers.
Classification: LCC TH1088.56 .L53 2023 | DDC 693.8/2--dc23/eng/20221206
LC record available at https://lccn.loc.gov/2022047189

ISBN: 978-1-032-26356-4 (hbk)
ISBN: 978-1-032-26357-1 (pbk)
ISBN: 978-1-003-28791-9 (ebk)

DOI: 10.1201/9781003287919

Typeset in Sabon
by SPi Technologies India Pvt Ltd (Straive)

Contents

Authors

Guo-Qiang Li is a professor of structural engineering at Tongji University, China, Director of the Research Centre of the Education Ministry of China for Steel Construction and Director of the National Research Centre of China for Pre-fabrication Construction.

Ling-Ling Wang is an associate professor of structural engineering at Huaqiao University, China.

Qing Xu is an assistant professor of structural engineering at the University of Shanghai for Science and Technology, China.

Jun-Wei Ge is a product manager at AkzoNobel Marine, Protective & Yacht Coatings in China.

Chapter 1

Introduction to intumescent coatings

1.1 DAMAGE OF STEEL STRUCTURES IN FIRE

Fire is one of the most common threats to public safety among all the various kinds of disasters that can occur. Fire disasters cause many thousands of deaths and huge economic loss each year. A bulletin (GA 2014) of the World Fire Statistics Center shows that costs due to losses from fire number in the tens of billions globally, and have been roughly estimated as approximately 1% of global GDP per annum. For Europe as a whole, the annual toll of fire deaths is measured in many thousands, with those suffering fire injuries numbered at many times more. In China, on average, more than 200,000 fires occur every year in which about 1,900 people are killed. During the period 1997–2017 this cost more than ¥2000 million in direct damage (Luo et al. 2021). A statistical report (NFPA 2021) of fire loss in the United States showed that more than one-third of fires occurred in or on structures and that most losses were caused by these fires.

Unprotected steel structures do not have the desired fire resistance. The temperature of unprotected steel elements increases rapidly in fire due to the high thermal conductivity of steel. In addition, the strength of steel is decreased with temperature elevation and drops to as low as around one-third of its original strength under 600°C (Li and Wang 2013, 45–48), as shown in Figure 1.1. The International Organization for Standardization (ISO) proposed a standard fire elevation (ISO 1975), the temperature of which may be increased over 600°C in 10 minutes and the maximum temperature of the fire may be up to 1000°C, as shown in Figure 1.2.

Since the load-bearing capacity of a steel structure, such as the steel beam shown in Figure 1.3, is proportional to the strength of the steel, the structure may eventually be damaged or even collapse when its capacity is decreased to a critical level which is not sufficient to carry the loads imposed on it (Figure 1.4). Therefore, steel structures are vulnerable to fire attack. Figures 1.5–1.7 are examples of steel structures severely damaged in fire (FEMA 2002; Intemac 2005).

DOI: 10.1201/9781003287919-1

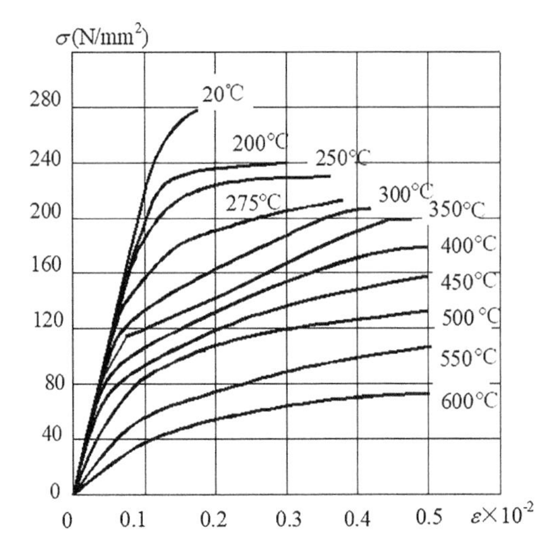

Figure 1.1 Strength reduction of steel with temperature elevation.

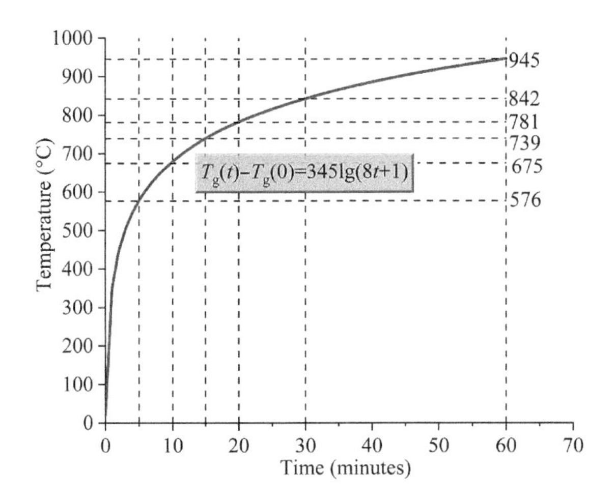

Figure 1.2 Temperature–time relationship of the ISO 834 standard fire.

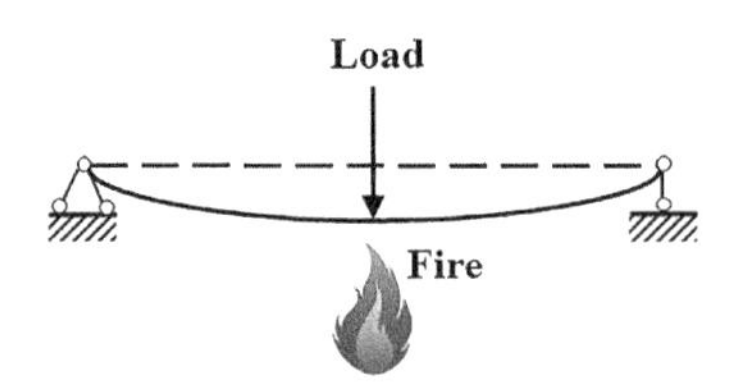

Figure 1.3 A steel beam supporting load under fire.

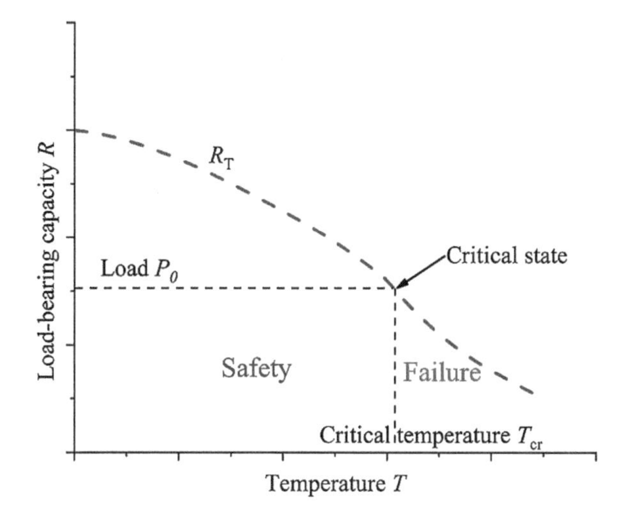

Figure 1.4 Capacity variation of a uniformly heated steel column without any restraint.

Figure 1.5 Collapse of a steel building in fire.

Figure 1.6 Severe distortion of a steel column in a World Trade Center building.

Figure 1.7 Collapse of a large part of the steel structure in the Windsor Tower Building.

1.2 TYPES OF INSULATIVE COATINGS FOR PROTECTING STEEL STRUCTURES AGAINST FIRE

1.2.1 Active fire protection and passive fire protection

There are two types of systems for building fire protection: active fire protection (AFP) and passive fire protection (PFP). It is important that both systems are properly working together in the event of a fire to ensure the safety of people in the building as well as the building itself. AFP uses systems that take action to put out the fire. PFP helps maintain or prolong the structural stability of the building subjected to fire, and also provides occupants with more time for evacuation.

AFP systems require some form of intervention or activation before they can work. These interventions may be manually operated, such as a fire extinguisher, or be automatic, such as a sprinkler (Figure 1.8). When fire and smoke are detected in a compartment of a building, a fire/smoke alarm will alert the sprinklers installed inside the building and work to actively put out or slow the growth of the fire. Sprinkler systems help slow the growth of the fire until firefighters arrive, who then use fire extinguishers and hoses to put out the fire altogether. AFP systems generally require routine maintenance and testing to ensure that they will work when required.

PFP covers a group of systems that compartmentalize a building through the use of fire-resistance-rated walls and floors, keeping the fire from spreading quickly and providing time for people to escape. Fire insulation of the primary structural elements of the building also falls under PFP. By a process of heat absorption and/or thermal insulation, PFP reduces the rate of temperature rise of the steel structures being protected. PFP systems require no intervention to work and if designed and chosen well will require little maintenance during the life of the installation.

Figure 1.8 Automatic sprinkler.

1.2.2 Non-intumescent and intumescent types of fire insulation for steel structures

Common types of fire insulation for protecting steel structures include non-intumescent coatings and boards, and intumescent coatings, all of which are applied to steel structural members such as beams and columns.

1.2.2.1 Non-intumescent coatings and boards

A non-intumescent type of fire insulation is commonly made of inorganic materials with low thermal conductivity, such as perlite, vermiculite and asbestos fibre, and mixed with cement and gypsum for adhesion. The final products of such insulation can be coatings or boards, which are applied to enclose steel elements with nailing, spraying or trowelling, as shown in Figures 1.9–1.11.

Non-intumescent coatings or boards do not react to heat in a fire. They form a good insulating layer to protect the steel substrates from heat.

Figure 1.9 Nailed board.

Figure 1.10 Sprayed coating.

Figure 1.11 Trowelled coating.

However, additional primers for corrosion protection purposes are normally required. Due to their loose structure, the mechanical characteristics, such as adhesion and strength, of non-intumescent coatings or boards are not good enough and easily damaged during service. In addition, the appearance of the coating is not smooth and good-looking, and is commonly used for the concealed components of steel structures.

1.2.2.2 Intumescent coatings

The intumescent type of fire insulation concerns the coatings that in the event of a fire react to heat by swelling in a controlled manner to 30 to 50 times of their original thickness to produce a carbonaceous char, which acts as an insulating layer to protect the steel substrates (Figure 1.12). The char creates insulation and reduces the rate of heating of the steel and extends its load bearing capacity. This type is normally applied by spray, brush, roller or trowel in the form of a liquid coating material that is applied onto a primer. Topcoats are normally applied onto intumescent films for the purpose of colour appearance, as well as protecting the films from water and moisture, as shown in Figure 1.13.

There are several types of intumescent coatings, including single-component water bases, single-component solvent bases and two-component epoxy or polyurethane bases. Single-component intumescent coatings are sensitive to water and moisture, and extra attention should be paid when using this type. Because of their good mechanical characteristics and architectural aesthetics, intumescent coatings have become more and more popular in application for protecting steel structures.

Figure 1.12 An intumescent coating reacts in fire to form a swelled char.

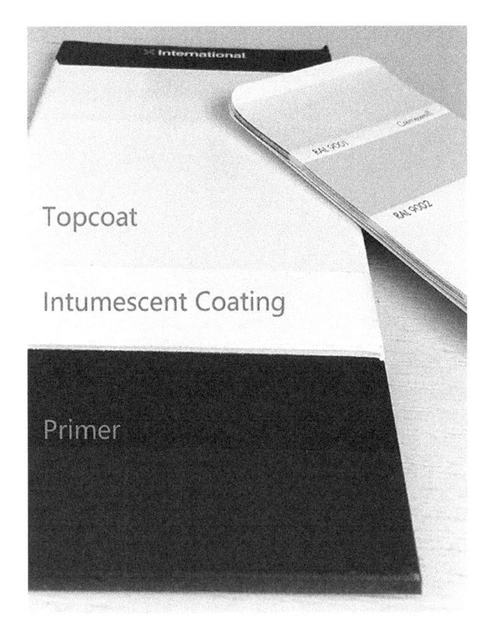

Figure 1.13 A steel panel coated with intumescent coating and topcoat.

1.3 APPLICATION OF INTUMESCENT COATINGS FOR PROTECTING STEEL STRUCTURES

1.3.1 Advantages of intumescent coatings

Thanks to the below listed advantages of intumescent coatings, more and more steel structure buildings are being protected by them.

1. The smooth and attractive appearance of the coatings can match the architectural aesthetics, especially for thin sized steel structures, as shown in Figures 1.14 and 1.15.
2. Because of their user-friendly features and good mechanical characteristics, they can be applied to steel members offsite in the workshop, which improves the construction quality and efficiency. Table 1.1 indicates the advantages of intumescent coatings by comparing them to non-intumescent cementitious coatings.
3. Since their good mechanical characteristics ensure the integrality of the coatings, particles hardly ever fall off them, which ensures a clean interior environment under the protected steel structures.

Figure 1.14 Steel column, beam and node coated with intumescent coating.

Figure 1.15 Facade supporting steel structures coated with tumescent coatings.

Figure 1.16 indicates the trends in the structural fire protection market in Great Britain, from 2001 to 2017 (Dolling 2018). The market share of off-site intumescent plus on-site intumescent coatings remains strong and the popularity of on-site intumescent coatings in the structural fire protection industry over the past 20 years has improved significantly. This has been driven by an intensely competitive industry which has in turn driven research and development.

1.3.2 Key points of application

Intumescent coatings are applied either onsite (in or adjacent to the building to be fire protected, either fully or partially erected) or offsite in a fabricator's shop with the fire protected steel then transported to the construction site and subsequently erected. Understanding the key points of application is critical in selecting the correct intumescent coating.

Table 1.1 Characteristics of intumescent coatings and non-intumescent coatings

Types of fire protection materials	*Application methods*	*Water resistance*	*Mechanical properties*	*Film appearance*	*Durability*
Intumescent single-component water base	By spray, brush, or roller	Very sensitive to water	Limited mechanical properties	Architectural aesthetics	Design life can be >15 years, but only for interior use
Intumescent single-component solvent base	By spray, brush, or roller	Sensitive to water	Limited mechanical properties	Architectural aesthetics	Design life can be >15 years, but only for interior use
Intumescent two-component solvent base	By spray or trowel	Good water resistance performance	Good mechanical properties	Architectural aesthetics	Design life can be >25 years for both interior and exterior use
Non-intumescent cementitious sprays	By spray or trowel	Sensitive to water	Poor mechanical properties	Appearance is very rough	Design life can be >15 years, but mainly for interior use

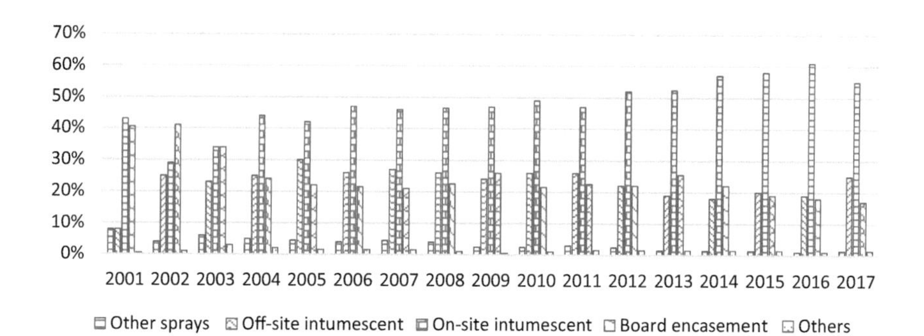

Figure 1.16 The market for the fire protection of steel frames: total market (beams + columns) by types of fire protection, Great Britain, 2001 to 2017.

1.3.2.1 Weather resistance

Intumescent coatings are formulated using special ingredients that, under the influence of heat, as in a fire, react and so create the intumescent expansion of the coating, forming an insulating layer on the steel. A disadvantage of these intumescent ingredients is that they tend to be relatively moisture sensitive. Unless adequately protected, intumescent coatings can therefore have only limited resistance to moisture and humid conditions which cause deterioration. So, if intumescent coatings are used improperly, the problem of bubbles on the coating film arises, as shown in Figure 1.17.

Intumescent coatings have different levels of resistance to weather depending upon the intumescent ingredients used and the resin platform the coating is based on. When referring to weather resistance we are primarily considering resistance to water in any form.

Generally, in order of resistance to weather, the most resistant are two-component formulated products, which successfully meet the most stringent requirements of environmental durability testing, followed by the single pack solvent-borne coatings, and then the single pack water-borne coatings.

Even within the single pack solvent-borne coatings there are formulation variants which offer better weather resistance than others. This is because some formulation is designed using specific grades of intumescent ingredients with lower water solubility but with the consequence of higher cost.

1.3.2.2 Using topcoats to improve water resistance

The topcoat to be used should be suitable for the service conditions that the system will be exposed to. Topcoats are typically applied at 50–75 microns

Figure 1.17 Bubbles developed on water-based intumescent coatings after raining.

dry film thickness (DFT) per coat, in one or two coats, with a total specified maximum thickness of 150 microns DFT. It is important that topcoats, when applied, form a continuous closed film and the minimum thickness is achieved. Failure to do this can result in early water ingress and damage of the intumescent coating.

For the less water-resistant intumescent coatings, it is recommended that the topcoat is applied as soon as the overcoating times allow and prior to likely exposure to adverse exterior conditions. For offsite application this means prior to moving coated steel work out of the shop and into an exterior environment. For application on-site this means prior to the coated steel work being exposed to precipitation and condensation conditions.

For intumescent grades specifically recommended for offsite application, provided the coating is not subject to continuous running water, standing/pooling water or immersion, it may remain without a topcoat for a period of up to six months. The exceptions to this recommendation are two-component formulated intumescent coatings which are designed, tested and proven for use in external exposure conditions without the need for topcoats.

Note that in extreme conditions topcoats will not offer full protection to the intumescent coating, for example, continuously running water, standing/pooling water and immersion.

1.3.2.3 Fire protection of connections

Fire protection of structural steelwork connections cannot be completed during offsite application of intumescent coatings. The applicator usually needs to make provision for the protection of bolts and possibly some areas of the connection onsite. This is completed after the steelwork is completely installed on site. The applicator should ensure the connection areas are properly masked during application, otherwise this could lead to issues related to connection joints and bolt tightening. Bolt caps provide an easy and efficient way of protecting connection onsite, as shown in Figure 1.18.

For welding connections, intumescent coatings and topcoats also cannot be completed during offsite application. After the steelwork is completely installed on site, proper surface preparation should be done before touching up primers. Intumescent coatings should be applied at the correct thickness before applying the topcoat.

1.3.2.4 Maintenance

It is important that a fire protection system is adequately maintained. The owner of a building should establish periodic inspection and the maintenance

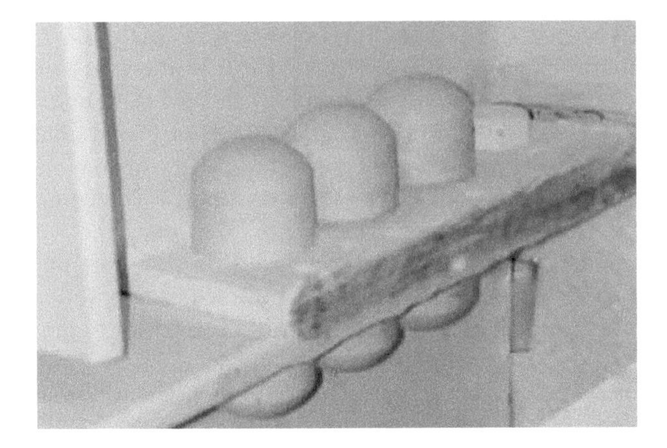

Figure 1.18 Bolt caps for fire protection on connection joints.

Figure 1.19 Degradation of the intumescent coating with a topcoat.

schedules of the fire protection system particularly where the exposure environment is more aggressive than ISO 12944 C1.

Any damage to the topcoat system will be at a point where moisture ingress is possible, leading to degradation of the intumescent coating, as shown in Figure 1.19. Any areas of damage should be repaired in line with the recommended procedures.

1.3.3 Typical steel structure projects protected with intumescent coatings

For architectural aesthetics purposes, there are many typical projects of steel structures protected with intumescent coatings, such as tall buildings, airport terminals, railway stations, exhibition centres, stadiums, conference centres, shopping malls, office buildings, manufacture plants and warehouses. Some good examples are introduced below.

1.3.3.1 Shanghai Tower building

Shanghai Tower stands in the heart of the Lujiazui Financial Zone in Shanghai Pudong New Area, as shown in Figure 1.20. The majestic high-rise boasts 127 floors above ground and five floors below ground. At 632 metres, Shanghai Tower is now the tallest building in China, and the second tallest in the world. The construction of Shanghai Tower started on November 29, 2008, and work was completed by the end of 2014. Trial operation was started in January 2017.

As shown in Figure 1.21, the curtain wall system of Shanghai Tower was designed as a symbiosis of two glazed walls – an exterior curtain wall (Curtain Wall A) and an interior curtain wall (Curtain Wall B) – with a tapering atrium in between. The main support for the exterior curtain wall

Figure 1.20 Shanghai Tower stands in the heart of the Lujiazui Financial Zone in Shanghai Pudong.

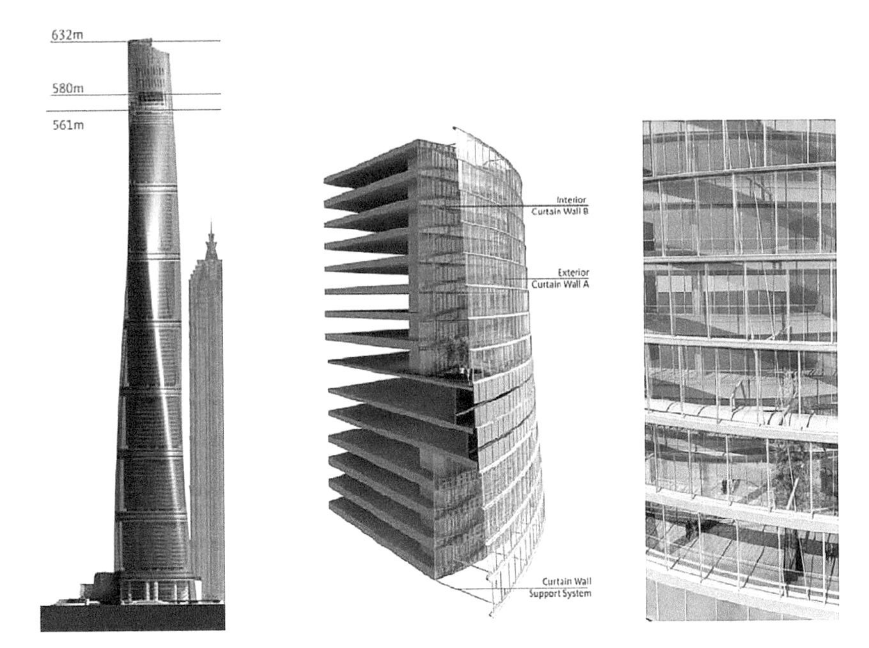

Figure 1.21 Curtain wall systems of Shanghai Tower.

is a horizontal ring beam consisting of a horizontal steel pipe 356 mm in diameter, laterally supported by a radial pipe strut (Sasha Zeljic 2010).

Special considerations were given to the fire protective requirements of the curtain wall support system. The solution was to use GB 14907 (AQSIQ 2002) tested for a 1.5 hour rating of intumescent coatings for the first three floors in each atrium. Considering the long external construction schedule and architectural aesthetics purpose, solvent-based intumescent coatings were used. A black topcoat was applied to the intumescent coatings to improve their water and humidity resistance, as shown in Figure 1.22.

1.3.3.2 Chengdu Tianfu New International Airport

Airport terminals, as typical large-span steel structural buildings, are one of the most common applications of intumescent coatings. A number of new airports have been built in China in recent years, of which Chengdu Tianfu New International Airport is an important one, as shown in Figure 1.23.

There are two terminals at this new airport, with a total construction area of 609,000 square metres. Over 300,000 square metres of steel beams and columns were coated with intumescent coatings for architectural aesthetics purposes. The fire ratings were 1.5 hours and 2.5 hours respectively, as shown in Figure 1.24.

Figure 1.22 Intumescent coating with topcoat on curtain wall support system in Shanghai Tower.

Figure 1.23 Chengdu Tianfu New International Airport.

Figure 1.24 Steel members of Chengdu Tianfu New International Airport protected with intumescent coatings.

1.3.3.3 Semiconductor manufacturing plants

The construction of semiconductor manufacturing plants has been blooming in recent years. To reduce the percentage of defective semiconductor manufacturing, such plants require a very clean interior environment, specified as a "clean room". For any construction material used in the clean room, its total level of molecular species off gassed in 30 min at 50°C should be less than 100 μg/g (Yan 2017). Cementitious sprays are unable to be used due to dust particles falling off. High quality intumescent coatings with ultra-low volatile organic compounds are a good choice.

An example of such a semiconductor manufacturing plant is the Huali Microelectronics Corporation (HLMC) 300 mm wafer foundry plant in Shanghai, as shown in Figure 1.25. The steel structural members of the plant are protected with intumescent coatings, as shown in Figure 1.26.

Figure 1.25 HLMC 300 mm wafer foundry plant in Shanghai, China.

Figure 1.26 Steel members of a semiconductor manufacturing plant protected with intumescent coatings.

1.4 MECHANISM OF INTUMESCENT COATINGS FOR FIRE PROTECTION OF STEEL STRUCTURES

In a fire, intumescent coatings react to heat by swelling in a controlled manner to 30 to 50+ times their original thickness to produce a carbonaceous char, which acts as an insulating layer to protect the steel substrates. The process is illustrated in Figure 1.27.

Figures 1.28 and 1.29 show a steel beam with an intumescent coating before and after a fire test respectively.

In a fire, the foaming agent melamine in the intumescent coating decomposes and releases ammonia, water, carbon dioxide, hydrogen halide and other non-flammable gases, which make the softened coating film foam to produce the carbonaceous char. The structure of such carbonaceous char is spongy, as shown in Figure 1.30. A number of pores are densely distributed in the spongy carbonaceous char, which makes the thermal conductivity low.

The carbonaceous char insulates heat, reducing the rate of heating of the steel and extends its load bearing capacity. The macroscopic thermal conductivity of the carbonaceous char is low, which insulates heat or reduces the rate of heating from fire to steel substrates, as shown in Figure 1.31. With such insulation, the temperature elevation of the protected steel is

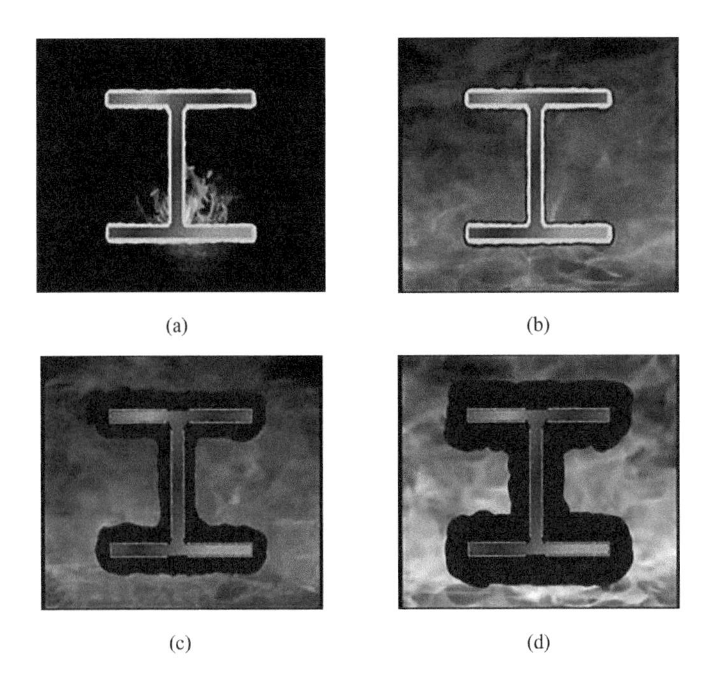

Figure 1.27 Variation of intumescent coating film in fire. (a) Fire starts to heat coating film. (b) Softened coating film starts to expand. (c) Film produces carbonaceous char. (d) Fully carbonaceous char is formed.

Figure 1.28 Steel beam with intumescent coating before fire test.

Figure 1.29 Steel beam with intumescent coating after fire test.

Figure 1.30 Spongy carbonaceous char.

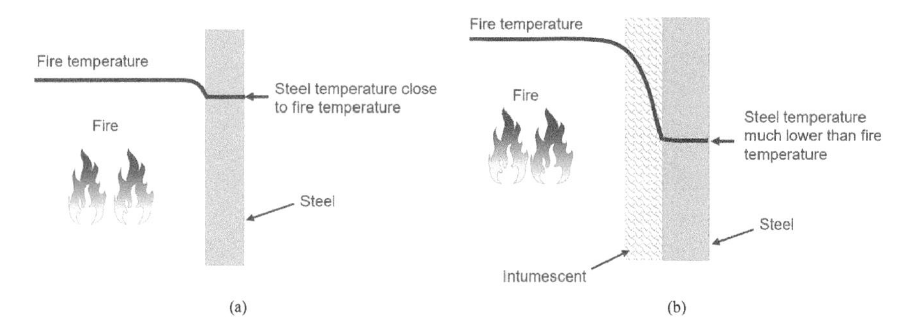

Figure 1.31 Comparison of heat transform to steel with and without fire protection. (a) Fire heating unprotected steel. (b) Fire heating protected steel.

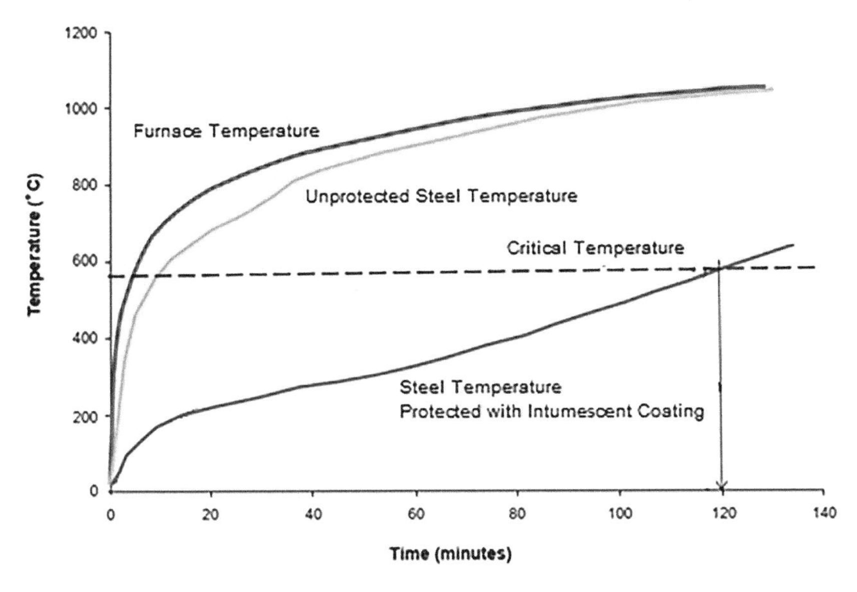

Figure 1.32 Temperature rises of steel with and without coating in fire.

significantly slowed down in comparison with the companion fire temperature, as shown in Figure 1.32.

1.5 MAIN ISSUES OF INTUMESCENT COATINGS FOR PROTECTING STEEL STRUCTURES

Due to the complexity of intumescent coatings, there are six major issues regarding protecting steel structures, which are very important in fire protection design.

1.5.1 Determining the thermal resistance of intumescent coatings

The temperature of the steel components needs to be evaluated beforehand when assessing the safety of steel building structures in fire. To enable calculation of the temperature of protected steel components in fire, the thermal resistance of the fire protection (insulation) material used has to be determined.

However, unlike conventional non-reactive fireproofing materials (e.g. concrete, gypsum, cementitious coatings), the intumescent coating is a thermally reactive fire protection material. The thickness and the insulation property of an intumescent coating change during the expansion process (Jimenez et al. 2006, 979). So, it is very hard to determine the thermal resistance of the intumescent coating, as the thickness and thermal conductivity of the coating changes in conditions of fire. Against this background, how to determine the thermal resistance of an intumescent coating is one of the most important issues to be solved.

1.5.2 Behaviour of intumescent coatings under large space fires

If fire occurs in small compartments, flashover will likely happen. The gas temperature in the compartments will elevate quickly after flashover and reach a very high level, even in excess of 1000°C. The fire temperature distribution in a small compartment may be considered to be approximately uniform. The temperature elevation curve of the ISO standard fire (ISO 1975) can be regarded as a representative of a small compartment fire.

However, if the fire takes place in a large space building, such as an airport terminal, a stadium or an exhibition centre, flashover will not happen. The fire will be localized at a specific area. The gas temperature in the large space is not uniformly distributed, but can be roughly divided into a hot air layer at the top of the space and a cool air layer at the bottom of the space (Li et al. 2018, 1), as shown in Figure 1.33.

The hot air at the top of a large space is actually the smoke of the fire. The temperature of such smoke is lower than the gas temperature of a small compartment fire after flashover, which can be regarded equivalently as the temperature of the fire's flames. The steel structures for the roof of a large space building will most likely meet such hot smoke induced by fire, which is called a large space fire in this book. A comparison of the temperature elevation of a large space fire with that of a small compartment fire and the ISO standard fire is shown in Figure 1.34.

As the chemical reactions of intumescent coatings rely on the heating condition of fire, the thermal behaviour of these coatings under a large space fire may be different from that under a small compartment fire, in particular, the representative of the ISO 834 standard fire. Under large space fires, intumescent coatings may not fully expand, whereas they may fully do so under a small compartment fire. So, it is necessary to identify the behaviour of

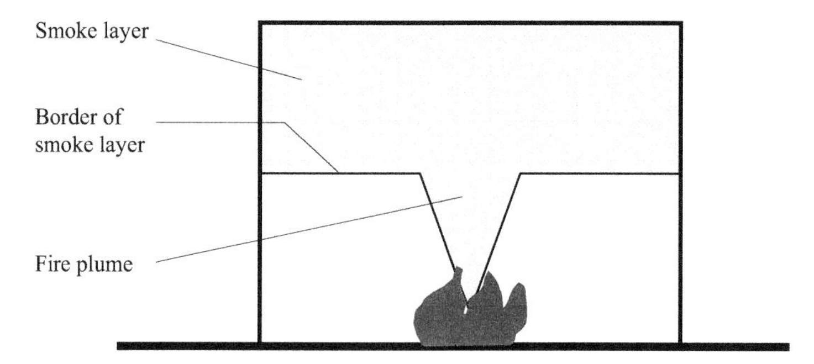

Figure 1.33 Illustration of a large space fire.

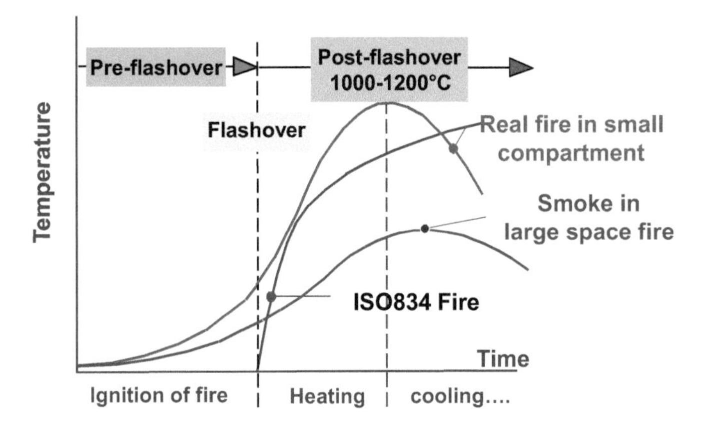

Figure 1.34 Comparison of a large space fire, a small compartment fire and the ISO standard fire.

intumescent coatings under a large space fire and the relationship with that behaviour under a small compartment fire.

1.5.3 Behaviour of intumescent coatings under localized fires

Steel components in a large space building are likely to be subjected to the localized fire flame when fire happens (Zhang et al. 2016, 62). However, the distribution of the gas temperature around the flame varies vastly: the closer to the flame, the higher the temperature. This fire condition is called a localized fire in this book, as shown in Figure 1.35.

Since the heating condition of protected steel members exposed to a localized fire varies throughout all the spots of steel members, which is different

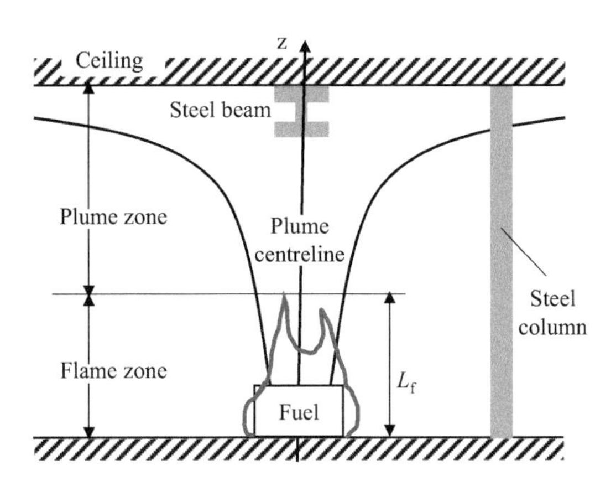

Figure 1.35 Illustration of a localized fire.

from the uniform gas temperature condition of a small compartment fire and a large space fire, the behaviour of intumescent coatings under such a localized fire needs special investigation.

1.5.4 Ageing effect of intumescent coatings

Since most of the chemical components in an intumescent coating are organic, exposure to long-term environmental conditions can cause the coating to lose some of its reactive materials, thus reducing the effectiveness of the coating over time. That the fire protection function of intumescent coatings deteriorates over time is called the ageing effect.

Among the service condition of intumescent coatings, solar radiation, temperature, water, moisture and a corrosive medium, such as salt, are the main factors inducing fire insulation degradation. Since the fire safety requirement needs to endure throughout the entire life of a steel structure, which may last 50 years or even longer, it is important to understand the long-term protection performance of an intumescent coating under exposure to different environmental conditions, which is the basis for the long-term fire protection design of steel structures.

1.5.5 Influence of the topcoats on the fire protection of intumescent coatings

Topcoats are often applied on the surface of intumescent coatings for protection against oxidation as well as for providing better appearance. They are considered indispensable for the long-term fire resistance performance of steel structures, particularly for outdoor applications where ageing is a major issue for intumescent coatings (Jimenez et al. 2016, A).

The influence of a topcoat on the fire protection of an intumescent coating is complex. Topcoats are very effective against UV and humidity attacks, which can reduce the decrease of a fire protection property. However, topcoats are generally thought to restrain the expansion process of the protected intumescent coatings, leading to lower expansion ratios and thus reduced insulation properties (Wang 2016, 143), even before ageing.

Most research on the durability of intumescent coatings indicates that the majority of samples used in ageing studies do not feature a topcoat. Understanding the influence of a topcoat on the fire protection of intumescent coatings is still one of the main issues for the fire safety of steel structures.

1.5.6 Temperature prediction of steel substrates protected by intumescent coatings

Since the mechanical properties of steel decrease rapidly with an increase in steel temperature, the precise prediction of steel component temperatures is most important for the fire safety evaluation of steel structures. Nowadays, the relationship of mechanical properties with the temperature of steel is available in fire safety design standards for steel structures, such as Chinese code GB 51249 (GB 2017, 19–20) and Eurocode EN 1994-1-2 (BSI 2005, 31–33); the capacity of steel components or even overall steel structures with elevated temperatures is also provided in these standards for evaluating structural safety under fire conditions. However, the availability of steel component temperatures in fire is the precondition for structural fire safety evaluation.

The method for calculating the temperature elevation of protected steel elements subjected to fire is provided in the codes based on the constant thermal resistance of insulative coatings. However, the thermal properties and thickness of intumescent coatings are extremely complex, and so a method for predicting the temperature elevation of steel elements protected with these coatings is not available. Therefore, seeking an accurate and reliable method to obtain the temperature of protected steel elements is also one of the key issues for evaluating the fire safety of steel structures with the protection of intumescent coatings.

The major issues regarding intumescent coatings for the fire protection of steel structures will be discussed in the following chapters.

REFERENCES

BSI. 2005. *Eurocode 4:Design of Composite Steel and Concrete Structures— Part 1–2: General Rules —Structural Fire Design.* BS EN 1994-1-2:2005. London: BSI.

Dolling, Chris. 2018. "The Market for Fire Protection of Steel Frames – Total Market (Beams + Columns) by Types of Fire Protection, Great Britain 2001 to 2017" Published at Steel Construction. Info website. https://steelconstruction.info. The market for fire protection of steel frames beams and columns png.

FEMA. 2002. *World Trade Center Building Performance Study: Date Collection, Preliminary Observation, and Recommendations.* Federal Emergency Management Agency. Washington DC

GA (Geneva Association). 2014. "Fire and Climate Risk." *Bulletin of World Fire Statistics, No. 29,* April, 2014.

GB. 2017. *Code for Fire Safety of Steel Structures in Buildings.* GB51249-2017. Beijing: China Planning Press.

Intemac (Instituto Tecnico De Materiales Y Construcciones). 2005. *Fire in the Windsor Building, Marid: Survey of the Fire Resistance and Residual Bearing Capacity of the Structure after the Fire.* Spain: Intemac.

ISO (International Organization for Standardization). 1975. *Fire Resistance Tests-Elements of Building Construction.* ISO 834: 1975. Geneva: ISO.

Jimenez, M., Bellayer, S., Naik, A., Bachelet, P., Duquesne, S., and Bourbigot, S. 2016. "Topcoats versus Durability of an Intumescent Coating." *Industrial and Engineering Chemistry Research* 55(36): 9625–9632.

Jimenez, M., Duquesne, S., and Bourbigot, S. 2006. "Characterization of the Performance of an Intumescent Fire Protective Coating." *Surface and Coating Technology* 201: 979–987.

Li, Guo-Qiang, and Wang, Pei-Jun. 2013. *Advanced Analysis and Design for Fire Safety of Steel Structures*. Berlin, Heidelberg: Springer-Verlag.

Li, Guo-Qiang, Xu, Qing, Han, Jun, Zhang, Chao, Lou, Guo-Biao, Jiang, Shou-Chao, and Jiang, Jian. 2018. "Fire Safety of Large Space Steel Buildings." *International Conference on Engineering Research and Practice for Steel Construction 2018 (ICSC2018)*, Hong Kong, China, 5 to 7 September.

Luo, Yi-Xin, Li, Qiang, Jiang, Lian-Rui, and Zhou, Yi-Hao. 2021. "Analysis of Chinese Fire Statistics during the Period 1997–2017." *Fire Safety Journal* 125(October): 103400.

NFPA (National Fire Protection Association). 2021. "Fire Loss in the United States During 2020." *NFPA Reports*, September, 2021.

Sasha Zeljic, Aleksandar. 2010. *Shanghai Tower Façade Design Process*. Published at Gensler company official website, https://www.gensler.com/doc/shanghai-tower-facade-deisgn-process-pdf

Wang, Ji. 2016. "The Protective Effects and Aging Process of the Topcoat of Intumescent Fire-Retardant Coatings Applied to Steel Structures." *Journal of Coatings Technology and Research* 13(1): 143–157.

Yan, Xu. 2017. "13 21 13 15 – Cleanroom Material Outgassing Requirements." M+W Group Design Specification: 7.

Zhang, Chao, Zhang, Zhe, and Li, Guo-Qiang. 2016. "Simple vs. Sophisticated Fire Models to Predict Performance of SHS Column in Localized Fire." *Journal of Constructional Steel Research* 120: 62–69.

Chapter 2

Determining the thermal resistance of intumescent coatings

For simply quantifying the thermal resistance of intumescent coatings for practical application, a comprehensive concept of constant effective thermal conductivity is proposed. This chapter introduces how to determine the constant effective thermal conductivity of intumescent coatings.

2.1 DEFINITION AND USAGE OF THE THERMAL CONDUCTIVITY OF MATERIALS

Thermal conductivity is a measure of a substance's ability to transfer heat through a material by conduction. Normally, materials with low thermal conductivity can be used as thermal insulation for steel protection against fire. According to the thermal equilibrium principle, the following differential equation can be established for a protected steel member subjected to fire:

$$c_s m_s \Delta T_s = \frac{\lambda\left(T_g - T_s\right) A_p}{d_p} \Delta t \tag{2.1}$$

where:

$$m_s = \rho_s V$$

A_p is the appropriate area of fire protection material per unit length of steel member (m^2/m);
V is the volume of steel member per unit length (m^3/m);
c_s is the temperature dependent specific heat of steel (J/kg·K);
c_p is the temperature dependent specific heat of the fire protection material (J/kg·K);
d_p is the thickness of the fire protection material (m);
m_s is the mass of the steel member (kg);

DOI: 10.1201/9781003287919-2

Δt is the time interval (s);
T_s is the steel temperature (°C);
T_g is the gas temperature during the fire (°C);
λ is the thermal conductivity of the fire protection material (W/m·K) ;
ρ_s is the unit mass of steel (kg/m³);
ρ_p is the unit mass of the fire protection material (kg/m³);
Δt is the time difference or interval (s);
ΔT_g is the increase of the fire gas temperature during the time interval Δt (K).

Equation (2.1) is a lumped capacitance heat transfer approximation to the transient heat conduction equation. The right side of the equation is the heat absorption by the steel member with its temperature increase, while the left side is the heat introduction to the steel member from fire getting through the fire insulation material in a specific interval.

Considering the heat loss effect by the absorption of the insulation material during the heat introduction from fire to the protected steel member, Eurocode EN 1993-1-2 (CEN 2005, 40) proposed a formula in incremental form to calculate the temperature of a steel member protected with insulative coatings, given by:

$$\Delta T_s = \frac{(T_g - T_s) A_p / V}{(d_p / \lambda) c_s \rho_s \left(1 + \frac{1}{3}\phi\right)} \Delta t - \left(e^{\phi/10} - 1\right) \Delta T_g \tag{2.2}$$

with:

$$\phi = \frac{c_p \rho_p d_p A_p}{c_a \rho_a V}$$

where:

A_p/V is the section factor of the steel member insulated with fire protection material.

The parameter, φ, is actually the ratio of the thermal capacity between the fire protection material and the steel member. When using intumescent coatings, the mass of the insulation material is very small compared to that of the protected steel member, so the thermal capacitance of intumescent coatings is negligible. Thus, Equation (2.2) can be simplified to:

$$\Delta T_s = \frac{A_p / V (T_g - T_s)}{(d_p / \lambda) c_s \rho_s} \Delta t \tag{2.3}$$

From the above equations, it can be seen that thermal conductivity is a fundamental parameter in the temperature prediction of protected steel structures in fire, which is needed for the assessment of structural fire safety.

When using the gas temperature of the ISO standard fire, a simple formula is provided in Chinese Code GB51249 (GB 2017, 26) for predicting the temperature elevation of the protected steel member as:

$$T_s = \left(\sqrt{0.044 + 5.0 \times 10^{-5} \frac{\lambda}{d_p} \frac{A_p}{V}} - 0.2 \right) t + 20 \quad (2.4)$$

where:

t is the specific duration of the fire (s).

2.2 THERMAL CONDUCTIVITY OF INTUMESCENT COATINGS

2.2.1 Time-dependent thermal conductivity

For many fire protection materials, the thermal conductivity could be approximately characterized as a fixed value. However, this is not the case for intumescent coatings. For determining an equivalent fixed value for intumescent coatings, the temperature-dependent thermal conductivity is firstly obtained by inverting the solution of Equations (2.3) and (2.4) as

$$\lambda = d_p c_s \rho_s \cdot \frac{V}{A_p} \cdot \frac{T_s(t+\Delta t) - T_s(t)}{\left[T_g(t+\Delta t) - T_s(t)\right]\Delta t} \quad (2.5a)$$

or

$$\lambda = \frac{\left((T_s - 20)/t + 0.2\right)^2 - 0.044}{5.0 \times 10^{-5}} \frac{d_p V}{A_p} \quad (2.5b)$$

2.2.2 Effective thermal conductivity

Obviously, calculating the temperature-dependent thermal conductivity by using Equation (2.5) requires the thickness of the intumescent coatings. However, this changes during fire and is very hard to measure directly (Jimenez et al. 2006, 979–987). To solve this problem, the concept of effective thermal conductivity is proposed to represent the thermal insulation property of intumescent coatings. The initial thickness of an intumescent coating, also called the dry film thickness (DFT), is employed for d_p in Equation (2.5) to determine the nominal effective thermal conductivity.

2.2.3 Constant effective thermal conductivity

Although the effective thermal conductivity provides reasonable results representing the insulation properties of intumescent coatings, it is temperature-dependent and difficult to use for engineers. For practical application, a constant effective thermal conductivity is proposed, which is defined as the average of effective thermal conductivities at protected steel temperatures in the range of 400–600°C, given by

$$\lambda_e = \frac{1}{200} \cdot \int_{T_s=400}^{T_s=600} \lambda(T_s)\,dT_s \tag{2.6}$$

where λ_e is the constant effective thermal conductivity (W/m K); and $\lambda(T_s)$ is the temperature-dependent thermal conductivity at the temperature of steel members, T_s (W/m K).

As most steel structures fail at temperatures in the range 400–600°C, the constant effective thermal conductivity may represent the insulation property of intumescent coatings to be used for the protected steel members at the high fire-induced temperatures of most interest. The benefit is that it can be employed to simply characterize the complex insulation property of intumescent coatings with a fixed constant value and which can also be easily utilized in predicting the temperature elevation of protected steel elements for fire safety evaluation.

2.3 TESTS FOR DETERMINING THE CONSTANT EFFECTIVE THERMAL CONDUCTIVITY OF INTUMESCENT COATINGS

In this section, a simple standard approach for testing the constant effective thermal conductivity of intumescent coatings is proposed (Li et al. 2016, 177–184; Li et al. 2017, 132–155).

2.3.1 Test specimens

The standard test specimen is a steel plate measuring 16 mm × 200 mm × 270 mm with thermocouples arranged as shown in Figure 2.1. For comparison, the H-shape section specimen and the channel-shape section specimen represent the commonly used steel elements are also tested. A 1 m long H-shape section specimen is shown in Figure 2.2. A total of 23 fire test specimens were carried out on the steel plate, the variable being the intumescent coating thickness. For fire tests on the H-shape and channel-shape section the specimens used were: nine 0.5 m long hot rolled H-shape ones with

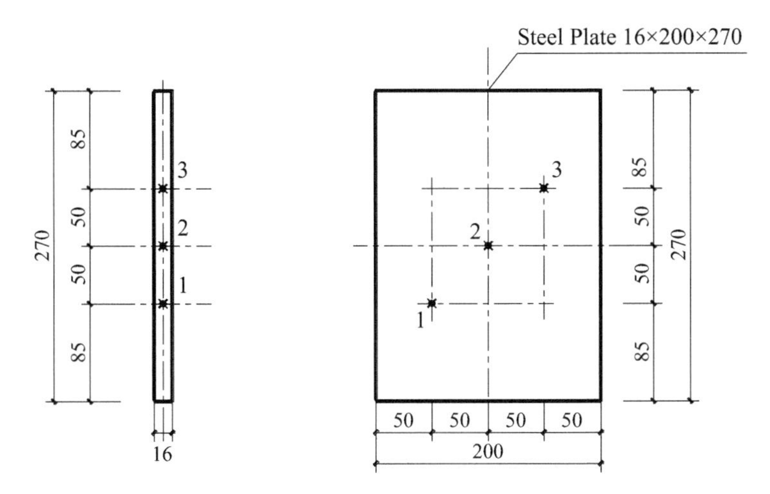

Figure 2.1 Dimensions of steel plate specimen and arrangement of thermocouples.

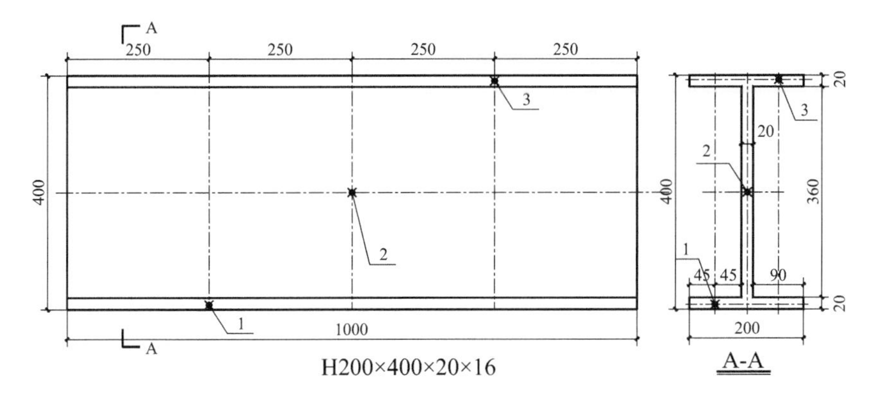

Figure 2.2 1 m long H-shaped section specimen and arrangement of thermocouples.

section 138 × 360 × 12 × 15.8; nine 0.5 m long hot rolled channel-shape ones with section 98 × 360 × 11 × 16; nine 1 m long welded H-shape ones with section 200 × 400 × 16 × 12; and nine 1 m long welded H-shape ones with section 200 × 400 × 20 × 16. The details for all the test specimens are listed in Table 2.1 (Han 2017, 47). The section factor of the steel specimens was in the range of (114.0, 155.3) m^{-1}.

One type of water based intumescent coating, denoted as 'T", and three types of solvent-based intumescent coating, denoted as 'L', 'A' and 'U' in Table 2.1, were used for the test specimens. The target DFT for type T intumescent coating were 0.3, 0.7, 2.2 and 2.9 mm. The target DFT of type L intumescent coating were 0.6, 0.9 and 1.5 mm. The target DFT of type A

Table 2.1 Details of all test specimens

Specimen ID*	Specimen type**	Section factor A_p/V (m^{-1})	Target DFT (mm)	Average measured DFT (mm)	Coating type
A03T1	Steel plate	125.0	0.3	0.323	T
A03T2	Steel plate	125.0	0.3	0.330	T
A07T1	Steel plate	125.0	0.7	0.761	T
A07T2	Steel plate	125.0	0.7	0.760	T
A07T3	Steel plate	125.0	0.7	0.711	T
A07T4	Steel plate	125.0	0.7	0.750	T
A07T5	Steel plate	125.0	0.7	0.701	T
A07T6	Steel plate	125.0	0.7	0.795	T
A22T1	Steel plate	125.0	2.2	2.245	T
A22T2	Steel plate	125.0	2.2	2.249	T
A22T3	Steel plate	125.0	2.2	2.250	T
A29T1	Steel plate	125.0	2.9	2.810	T
A29T2	Steel plate	125.0	2.9	3.110	T
A29T3	Steel plate	125.0	2.9	3.085	T
B03T1	I-shaped section	142.1	0.3	0.296	T
B07T1	I-shaped section	142.1	0.7	0.792	T
B07T2	I-shaped section	142.1	0.7	0.802	T
B07T3	I-shaped section	142.1	0.7	0.761	T
B07T4	I-shaped section	142.1	0.7	0.818	T
B07T5	I-shaped section	142.1	0.7	0.797	T
B22T1	I-shaped section	142.1	2.2	2.210	T
B22T2	I-shaped section	142.1	2.2	2.340	T
B29T1	I-shaped section	142.1	2.9	2.932	T
C03T1	C-shaped section	155.3	0.3	0.282	T
C07T1	C-shaped section	155.3	0.7	0.709	T
C07T2	C-shaped section	155.3	0.7	0.763	T
C07T3	C-shaped section	155.3	0.7	0.817	T
C07T4	C-shaped section	155.3	0.7	0.790	T
C07T5	C-shaped section	155.3	0.7	0.803	T
C22T1	C-shaped section	155.3	2.2	2.156	T
C22T2	C-shaped section	155.3	2.2	2.071	T
C29T1	C-shaped section	155.3	2.9	2.876	T
D06L1	Steel plate	142.5	0.6	0.626	L
D09L1	Steel plate	142.5	0.9	0.897	L
D15L1	Steel plate	142.5	1.5	1.548	L
E06L1	H-shaped section	145.7	0.6	0.620	T
E06L2	H-shaped section	145.7	0.6	0.626	T
E06L3	H-shaped section	145.7	0.6	0.613	T
E09L1	H-shaped section	145.7	0.9	0.920	T

(Continued)

Table 2.1 (Continued) Details of all test specimens

*Specimen ID**	*Specimen type***	*Section factor A_p/V (m^{-1})*	*Target DFT (mm)*	*Average measured DFT (mm)*	*Coating type*
E09L2	H-shaped section	145.7	0.9	0.856	T
E09L3	H-shaped section	145.7	0.9	0.872	T
E15L1	H-shaped section	145.7	1.5	1.476	T
E15L2	H-shaped section	145.7	1.5	1.577	T
E15L3	H-shaped section	145.7	1.5	1.483	T
F06L1	H-shaped section	114.0	0.6	0.674	T
F06L2	H-shaped section	114.0	0.6	0.607	T
F06L3	H-shaped section	114.0	0.6	0.571	T
F09L1	H-shaped section	114.0	0.9	0.804	T
F09L2	H-shaped section	114.0	0.9	0.804	T
F09L3	H-shaped section	114.0	0.9	0.807	T
F15L1	H-shaped section	114.0	1.5	1.262	T
F15L2	H-shaped section	114.0	1.5	1.337	T
F15L3	H-shaped section	114.0	1.5	1.309	T
AZ-1-00-1	Steel plate	125.0	1	0.930	A
AZ-1-00-2	Steel plate	125.0	1	1.080	A
AZ-1-00-3	Steel plate	125.0	1	1.050	A
AZ-2-00-1	Steel plate	125.0	2	2.180	A
AZ-2-00-2	Steel plate	125.0	2	2.230	A
AZ-2-00-3	Steel plate	125.0	2	2.190	A
UI-1-00-1	Steel plate	125.0	1	0.900	U
UI-1-00-2	Steel plate	125.0	1	0.910	U
UI-1-00-3	Steel plate	125.0	1	1.040	U

* Specimen ID = specimen type** + target DFT + type of coating + sample number;
** The notations for specimen type include: A/AZ/UI for steel plate, B for hot rolled I-shaped section, C for hot rolled channel-shaped section, D for steel plate section (full protected with intumescent coating), E for welding H-shaped section-1, and F for welding H-shaped section-2.

intumescent coating were 1.0 and 2.0 mm. The target DFT of type U intumescent coating were 1.0 mm.

The test specimens with applied intumescent coating, as shown in Figure 2.3, were dried at ambient temperature for at least 48 h until thoroughly dried. The DFTs of each specimen were measured at a number of locations using a digital thickness gauge; the average DFTs are listed in Table 2.1.

The steel plates for type T intumescent coating were insulated around the edges. The steel plates for type L intumescent coating were protected by the same intumescent coating and the plates were exposed to fire all round. This resulted in slightly different section factors for these two types of intumescent coatings, as shown in Table 2.1.

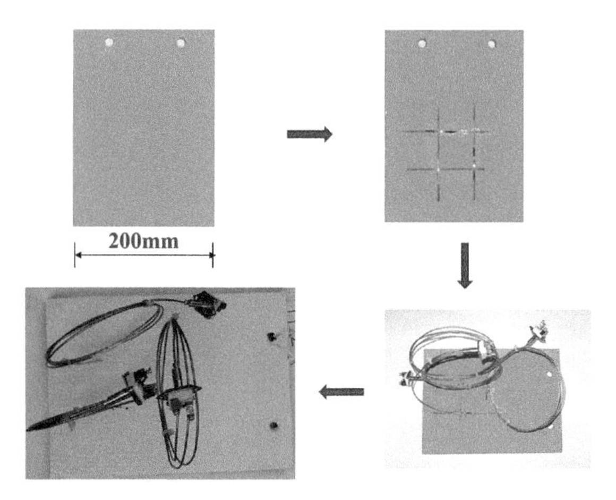

Figure 2.3 Test specimens with applied intumescent coating.

2.3.2 Test setup

Two fire testing furnaces of Tongji University were used for the fire tests listed in Table 2.2: a small furnace (internal dimensions 1000 mm × 1000 mm × 1200 mm) as shown in Figures 2.4 and 2.5, and a large one (internal dimensions 4500 mm × 3000 mm × 2200 mm), as shown in Figures 2.6 and 2.7. The interior surfaces of the furnaces were lined with ceramic fibre

Table 2.2 Summary of fire tests

Test no.	*Test furnace*	*Test specimens*
1	Small	A03T1/A03T2/B03T1/C03T1
Test 2	Small	A07T1/A07T2/A07T3
Test 3	Small	A07T4/A07T5/A07T6
Test 4	Small	A22T1/A22T2/A22T3
Test 5	Small	A29T1/A29T2/A29T3
Test 6	Small	B07T1/B07T2/B07T3
Test 7	Small	B22T1/B22T2/B29T1
Test 8	Small	C07T1/C07T2/C07T3
Test 9	Small	C22T1/C22T2/C29T1
Test 10	Small	B07T4/B07T5/C07T4/C07T5
Test 11	Small	D06L1/D09L1/D15L1
Test 12	Large	E05L1/E05L2/E05L3/E10L1/E10L2/E10L3/F05L1/F05L2/F05L3
Test 13	Large	E15L1/E15L2/E15L3/F10L1/F10L2/F10L3/F15L1/F15L2/F15L3
Test 14	Small	AZ-1-00-1/AZ-1-00-2/AZ-1-00-3/AZ-2-00-1/AZ-2-00-2/ AZ-2-00-3/UI-1-00-1/UI-1-00-2/UI-1-00-3/

Figure 2.4 Small furnace of Tongji University.

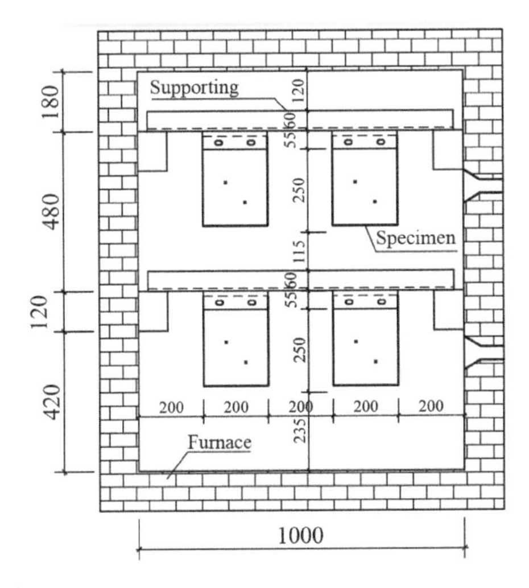

Figure 2.5 Cross-section dimensions of the small furnace and arrangement of steel plate test specimens.

materials of thickness 200 mm to efficiently transfer heat to the specimens. The small furnace was used for fire tests on all the steel plate specimens, and the hot rolled H-shape and channel-shape section specimens, while the large furnace was used for testing all the welded H-shape section specimens.

The fire temperature–time evolution in both furnaces was controlled according to the ISO 834 standard fire condition (ISO 2015, 12).

Figure 2.6 Large fire test furnace of Tongji University.

Figure 2.7 Positioning of steel section specimen in the large furnace.

2.3.3 Test results and discussions

2.3.3.1 Furnace and steel temperatures

The typical gas temperatures recorded at different locations in the large furnace (Test 13) indicates nearly uniform temperature distribution in the even relatively large space of the furnace, as shown in Figure 2.8. The average gas temperatures of all the fire tests are shown in Figure 2.9, confirming that all the fire tests almost achieved the intended standard fire temperature–time curve.

The typical steel temperatures measured at different locations are shown in Figure 2.10, revealing uniform temperature distribution in the steel section.

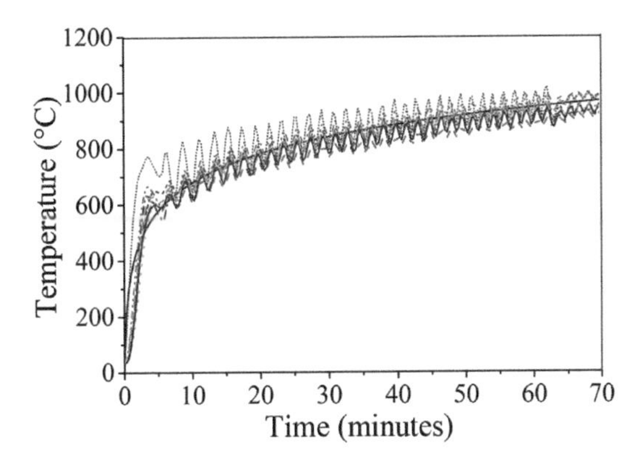

Figure 2.8 Gas temperature–time relationships measured inside the large furnace (Test 13).

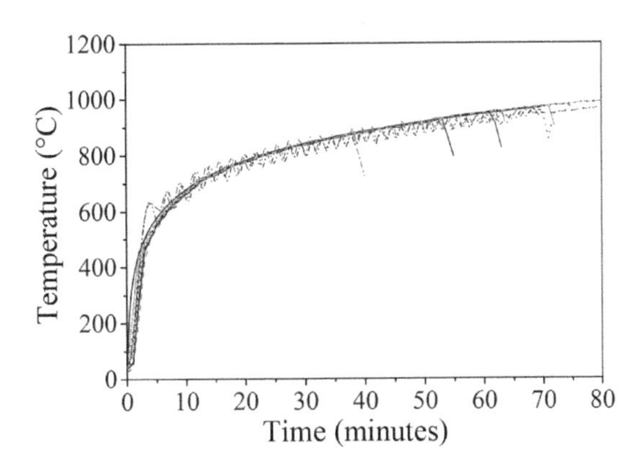

Figure 2.9 Average furnace gas temperature–time relationships for all tests.

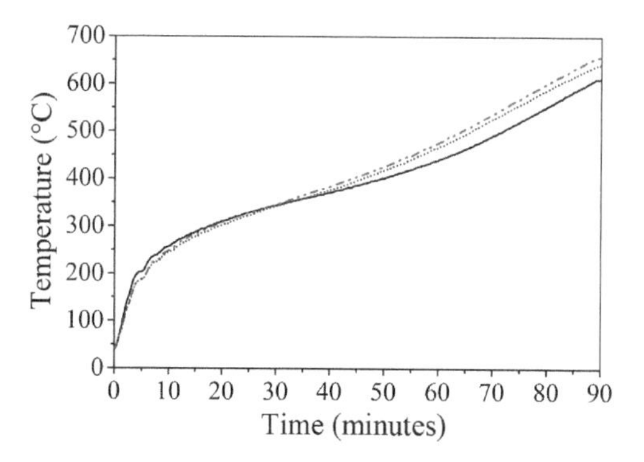

Figure 2.10 Typical steel temperatures measured at different locations (Test E15L3).

2.3.3.2 Effective thermal conductivity

The intumescent coatings applied on the test specimens reacted and expanded when exposed to fire in the furnace, as shown in Figure 2.11 (Wang et al. 2013, 171). The variation of the effective thermal conductivity of the coatings with coating temperatures obtained from the tests is illustrated in Figure 2.12, in which the intumescent coating temperature was taken as the mean value of the steel and fire temperatures.

It can be seen from Figure 2.12 that the effective thermal conductivity of the intumescent coatings started to fall sharply after their temperature reached about 100°C, indicating chemical reactions starting around that temperature. The effective thermal conductivity of the coatings was reduced

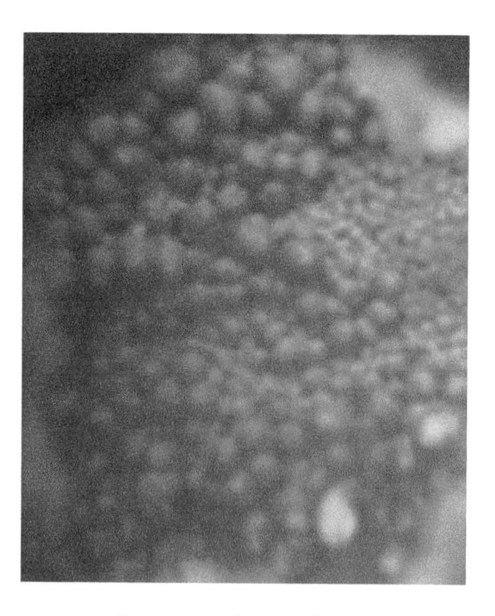

Figure 2.11 Intumescent coating expanded in fire tests.

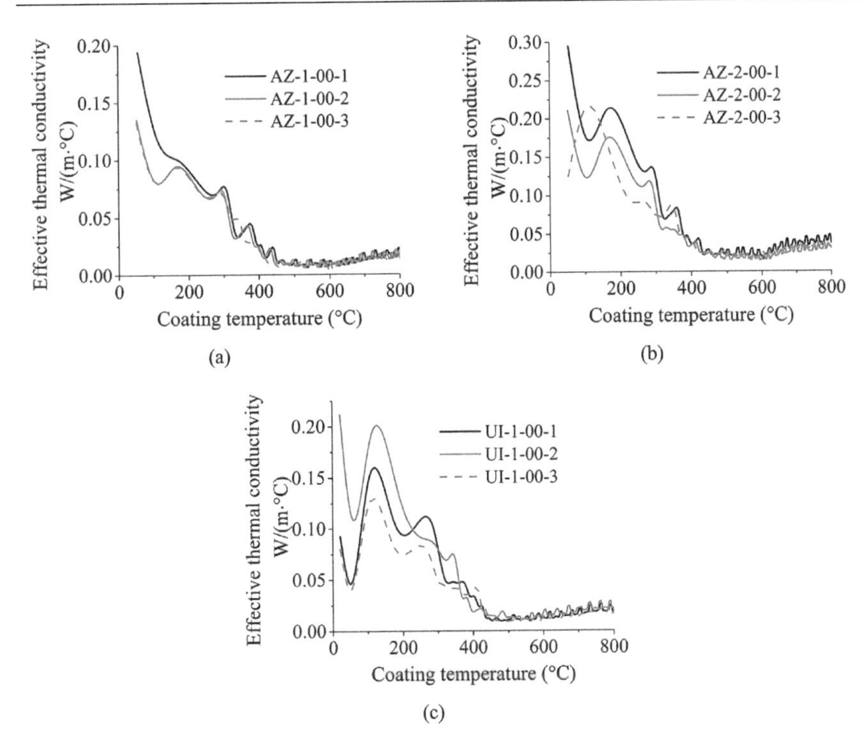

Figure 2.12 Variation of effective thermal conductivity with coating temperatures. (a) Type AZ-1 specimens. (b) Type AZ-2 specimens. (c) Type U specimens.

to a very low value at about 400°C, which is a clear indication that the coating had reached full expansion. When the effective thermal conductivity reached the minimum at around 450–500°C, it slightly increased with greater elevated temperatures, which may be explained by increased radiation inside the bubbles created in the intumescent coatings (Yuan 2009, 126).

2.3.3.3 Constant effective thermal conductivity

The constant effective thermal conductivity values obtained for all the different tests is presented in Table 2.3.

The raw temperature-dependent thermal conductivity using Equation (2.5) with the constant effective thermal conductivity for one of the tests (Test 7 in Table 2.2) are compared in Figure 2.13. It can be seen that the raw thermal conductivity decreases drastically in the initial phase of fire during the expanding stage. After full expansion, the thermal conductivity stabilizes. The constant effective thermal conductivity, which is temperature independent, calculated using Equation (2.6), represents the average of thermal conductivities during the stable stage. This is sensible because it is the

Table 2.3 Constant effective thermal conductivities

Specimen ID	*Measured (W/m·K)*	*Average (W/m·K)*
A03T1	0.0092	0.0093
A03T2	0.0093	
A07T1	0.0178	
A07T2	0.0157	
A07T3	0.0145	0.0159
A07T4	0.0155	
A07T5	0.0160	
A07T6	0.0158	
A22T1	0.0400	
A22T2	0.0394	0.0434
A22T3	0.0508	
A29T1	0.0386	
A29T2	0.0411	0.0407
A29T3	0.0424	
B03T1	0.0093	0.0093
B07T1	0.0089	
B07T2	0.0091	
B07T3	0.0106	0.0103
B07T4	0.0115	
B07T5	0.0114	
B22T1	0.0216	
B22T2	0.0208	0.0212
B29T1	0.0239	0.0239
C03T1	0.0075	0.0075
C07T1	0.0079	
C07T2	0.0094	
C07T3	0.0088	0.0089
C07T4	0.0089	
C07T5	0.0095	
C22T1	0.0210	
C22T2	0.0178	0.0194
C29T1	0.0187	0.0187
D06L1	0.0063	0.0063
D09L1	0.0089	0.0089
D15L1	0.0099	0.0099
E06L1	0.0066	
E06L2	0.0063	0.0065
E06L3	0.0065	
E09L1	0.0079	
E09L2	0.0073	0.0076

(Continued)

Table 2.3 (Continued) Constant effective thermal conductivities

Specimen ID	*Measured (W/m·K)*	*Average (W/m·K)*
E09L3	0.0075	
E15L1	0.0103	
E15L2	0.0104	0.0101
E15L3	0.0096	
F06L1	0.0103	
F06L2	0.0096	0.0096
F06L3	0.0089	
F09L1	0.0101	
F09L2	0.0093	0.0099
F09L3	0.0102	
F15L1	0.0122	
F15L2	0.0130	0.0128
F15L3	0.0132	

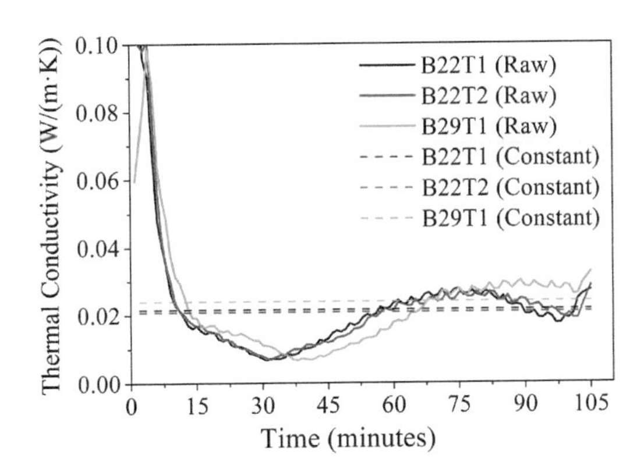

Figure 2.13 Comparison between raw temperature-dependent thermal conductivity and constant effective thermal conductivity (Test 7).

expanded intumescent char that provides effective fire protection to the steel substrates.

The duplicate tests for each arrangement (same section size, same coating thickness, and same intumescent coating type) produced very similar results, as shown in Table 2.3. Therefore, in the discussions below, the average constant effective thermal conductivity is used.

2.3.3.4 *Comparison of constant effective thermal conductivity between various specimens*

The comparison of constant effective thermal conductivity values between various specimens are shown in Figure 2.14, grouped according to the intumescent coating thickness and steel section factor. It can be seen that the effects of the intumescent coating thickness and steel section factor on the constant effective thermal conductivity of Type L intumescent coating are not apparent, while the effect of thickness is much greater than that of steel section factors on the constant effective thermal conductivity of Type T intumescent coating. These results suggest that it would be possible to use steel plate specimens as a substitute for more realistic steel member specimens with I-shaped and C-shaped sections for fire tests to obtain the constant effective thermal conductivity of intumescent coatings.

The results in Figure 2.14 also indicate that the constant effective thermal conductivity of intumescent coatings tends to increase with increasing DFT, which means that the constant thermal insulative efficiency of intumescent coatings decreases with the decreasing rate of heat transformation. Hence, it is proposed that the constant effective thermal conductivity of intumescent coatings needs evaluation through fire tests on protected steel specimens with a specific thickness of coatings.

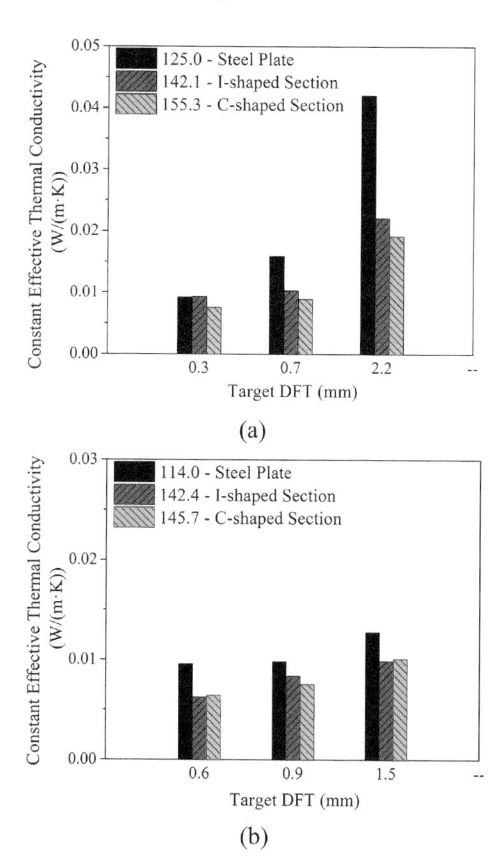

Figure 2.14 Effects of steel section factor and DFT on constant effective thermal conductivity of intumescent coatings. (a) Type T coating. (b) Type L coating.

2.4 USE OF CONSTANT EFFECTIVE THERMAL CONDUCTIVITIES FOR INTUMESCENT COATINGS

The constant effective thermal conductivities for intumescent coatings can be used to calculate the protected steel elements exposed to fire. The average constant effective thermal conductivity values in Table 2.3 have been used to calculate steel temperatures for the fire tested steel plates and steel members with various sections. The measured average and calculated steel temperatures are compared in Figure 2.15. It can be seen that, in most cases,

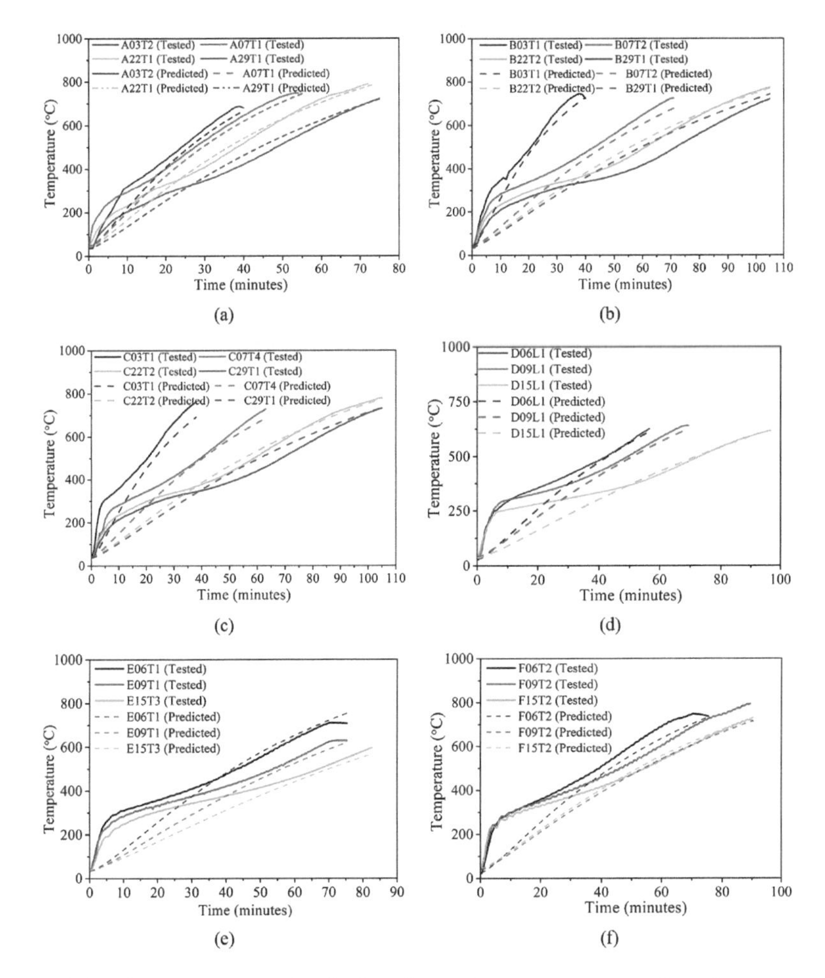

Figure 2.15 Comparisons of calculated steel temperatures using the constant effective thermal conductivity with experimental results. (a) Steel plate with Type T coating. (b) I-section specimen with Type T coating. (c) C-section specimen with Type T coating. (d) Steel plate with Type L coating. (e) H-section specimens 1 with Type L coating. (f) H-section specimens 2 with Type L coating.

the calculated results agree with those measured from the tests very well. Because the constant effective thermal conductivity was determined based on averaging the temperature dependent thermal conductivities within the steel temperature range of 400–600°C, the agreement between the calculated and measured steel temperatures within this range is excellent in all cases. In effect, because the thermal conductivity of intumescent coating changes slowly when the steel temperature exceeds 400°C, as shown in Figure 2.12, the calculated steel temperatures using a constant effective thermal conductivity higher than 600°C are also very close to the test results. This can be taken as a demonstration of the feasibility of using a constant characterized value as a representative for thermal conductivity of intumescent coatings.

2.5 SUMMARY

In this chapter, the thermal properties of intumescent coatings are discussed. It has been found that the thermal conductivity of intumescent coatings is a temperature-dependent variable, while the variation of the effective thermal conductivity is stabilized after full expansion when the temperature of the protected steel substance exceeds 400°C. The concept of constant effective thermal conductivity has been introduced with the average of the effective thermal conductivities in the range of 400 to 600°C as a characterized representative of the insulation property of intumescent coatings. A simple standard test approach has been proposed for quantifying the constant effective thermal conductivity of intumescent coatings with various DFTs.

Based on the test results, these useful findings may be drawn:

1. The intumescent coatings begin to react to heat and expand at around 100°C, with complete full expansion at about 400°C.
2. The constant effective thermal conductivity of intumescent coatings increases with the increase of the coating thickness.
3. The constant effective thermal conductivities tested from the steel plate specimens and the specimens with other steel sections are similar.

It has also been found that the temperature over 400°C of steel elements protected with intumescent coatings can be satisfactorily predicted with the corresponding constant effective thermal conductivity in the condition of full expansion.

REFERENCES

CEN. 2005. *EN 1993-1-2:2005.Eurocode 3: Design of Steel Structures — Part 1–2: General Rules — Structural Fire Design*. Brussels: European Committee for Standardization.

GB. 2017. *GB51249-2017. Code for Fire Safety of Steel Structures in Buildings*. Beijing: China Planning Press.

Han. Jun. 2017. *Study on the Performance of Intumescent Coating under Large Space Fires*. Shanghai: Tongji Unviersity.

ISO. 2015. *ISO 834–1: 2015, Fire-Resistance Tests — Elements of Building Construction —Part 1: General Requirements*. Geneva, Switzerland: ISO.

Jimenez, M., Duquesne, S., and Bourbigot, S. 2006. "Characterization of the Performance of an Intumescent Fire Protective Coating." *Surface and Coating Technology* 201: 979–987.

Li, Guo-Qiang, Han, Jun, Lou, Guo-Biao, and Wang, Yong C.. 2016. "Predicting Intumescent Coating Protected Steel Temperature in Fire Using Constant Thermal Conductivity." *Thin-Walled Structures* 98: 177–184. https://doi.org/10.1016/j.tws.2015.03.008.

Li, Guo-Qiang, Han, Jun, and Wang, Yong C. 2017. "Constant Effective Thermal Conductivity of Intumescent Coatings: Analysis of Experimental Results." *Journal of Fire Sciences* 35(2): 132–155. https://doi.org/10.1177/0734904117693857.

Wang, Ling-Ling, Wang, Yong C., and Li, Guo-Qiang. 2013. "Experimental Study of Hydrothermal Aging Effects on Insulative Properties of Intumescent Coating for Steel Elements." *Fire Safety Journal* 55: 168–181. https://doi.org/10.1016/j.firesaf.2012.10.004.

Yuan, Ji-Feng. 2009. *Intumescent Coating Performance on Steel Structures under Realistic Fire Conditions*. Manchester: University of Manchester.

Chapter 3

Behaviour of intumescent coatings under large space fires

Since the temperature elevation of a large space fire is lower than that of a post-flashover fire or the standard fire (Li et al. 2018, 1; Li and Wang 2013, 22–27), the behaviour of intumescent coatings exposed to a large space fire should be different (Yuan 2009; Wang et al. 2015, 627–643; Lucherini et al. 2018, 42–50). To understand this difference, an experimental investigation on the behaviour of intumescent coatings subjected to a large space fire was conducted in comparison with that of the ISO 834 standard fire.

3.1 FIRE CONDITIONS

The ISO 834 standard fire and three large space fire temperature–time curves were employed for the experimental investigation, denoted as T0 and T1–T3 respectively. The temperature–time curves of the large space fires accorded with Chinese technical code CECS 200 (CECS 2006, 18) and were compared with that of the ISO 834 standard fire curve, as shown in Figure 3.1. The enclosure space dimensions associated with heat release rates (HRRs) for the three different large space fires (Du 2007, 15–31) are listed in Table 3.1.

3.2 TEST PREPARATION

3.2.1 Test specimens

A total of 54 specimens were tested on intumescent-coating-protected steel plate specimens, which measured 5 mm × 300 mm × 300 mm. The section factor was 200 m^{-1} for the steel plate specimens as only one side was exposed to fire. Three type-K thermocouples were attached to each specimen as shown in Figure 3.2 (Han 2017, 96).

The test variables included the intumescent coating's dry film thickness (DFT), the fire exposure condition and the smoke condition. Table 3.2 lists the main test variables, in which T0 to T3 represent the different fire temperature–time curves, including the ISO standard fire curve (T0) and the

DOI: 10.1201/9781003287919-3

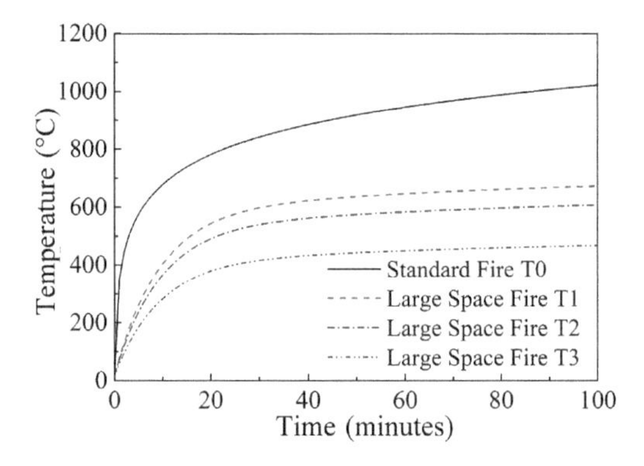

Figure 3.1 Comparison between measured furnace temperature–time curves (dashed lines) and the expected (solid line) under various fire conditions.

Table 3.1 Parameters of the three large space fires

Fire conditions	*Typical enclosure dimensions*	*Maximum gas temperature (°C)*	*HRR (MW)*
Fire T1	Area 1000 m^2; height 6 m	700	>15
Fire T2	Area 3000 m^2; height 6 m	630	>15
Fire T3	Area 3000 m^2; height 12 m	480	>15

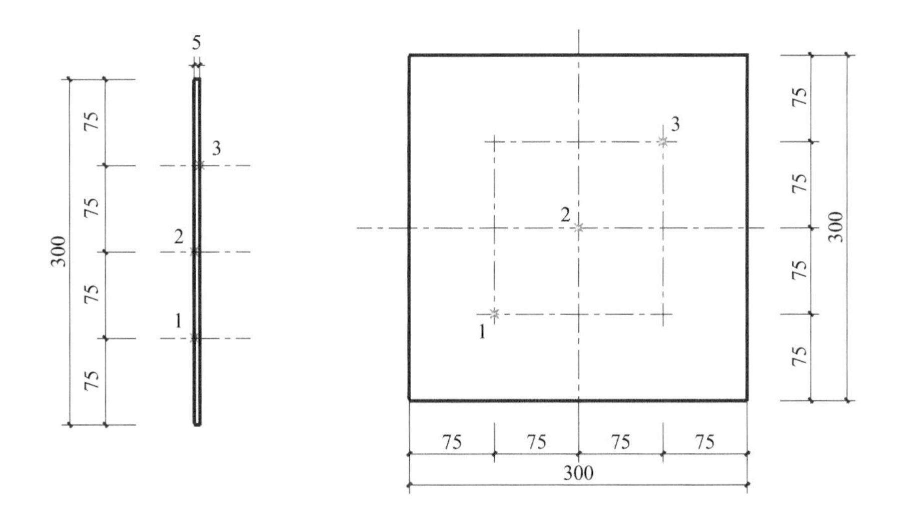

Figure 3.2 Dimensions of steel plate specimens and arrangement of thermocouples.

Table 3.2 Main variables of specimens for tests

*Specimen ID**	*Fire condition*	*Target DFT (mm)*	*Average measured DFT (mm)*	*Duplicate specimen ID**	*Average measured DFT (mm)*
T0S0L1N1	T0	0.37	0.385	T0S0L1N2	0.404
T0S0L2N1	T0	0.60	0.657	T0S0L2N2	0.669
T0S0L3N1	T0	1.29	1.327	T0S0L3N2	1.306
T0S1L1N1	T0	0.37	0.400	T0S1L1N2	0.407
T0S1L2N1	T0	0.60	0.674	T0S1L2N2	0.668
T0S1L3N1	T0	1.29	1.300	T0S1L3N2	1.339
T0S2L1N1	T0	0.37	0.414	T0S2L1N2	0.408
T0S2L2N1	T0	0.60	0.660	T0S2L2N2	0.676
T0S2L3N1	T0	1.29	1.304	T0S2L3N2	1.295
T1S0L1N1	T1	0.37	0.397	T1S0L1N2	0.405
T1S0L2N1	T1	0.60	0.646	T1S0L2N2	0.654
T1S0L3N1	T1	1.29	1.309	T1S0L3N2	1.328
T1S1L1N1	T1	0.37	0.400	T1S1L1N2	0.406
T1S1L2N1	T1	0.60	0.686	T1S1L2N2	0.681
T1S1L3N1	T1	1.29	1.290	T1S1L3N2	1.302
T1S2L1N1	T1	0.37	0.409	T1S2L1N2	0.401
T1S2L2N1	T1	0.60	0.681	T1S2L2N2	0.638
T1S2L3N1	T1	1.29	1.317	T1S2L3N2	1.316
T2S0L1N1	T2	0.37	0.407	T2S0L1N2	0.408
T2S0L2N1	T2	0.60	0.660	T2S0L2N2	0.674
T2S0L3N1	T2	1.29	1.273	T2S0L3N2	1.283
T2S1L1N1	T2	0.37	0.394	T2S1L1N2	0.403
T2S1L2N1	T2	0.60	0.676	T2S1L2N2	0.680
T2S1L3N1	T2	1.29	1.328	T2S1L3N2	1.328
T2S2L1N1	T2	0.37	—	T2S2L1N2	—
T2S2L2N1	T2	0.60	0.686	T2S2L2N2	0.675
T2S2L3N1	T2	1.29	1.291	T2S2L3N2	1.280
T3S0L1N1	T3	0.37	0.403	T3S0L1N2	0.404
T3S0L2N1	T3	0.60	0.693	T3S0L2N2	0.664
T3S0L3N1	T3	1.29	1.304	T3S0L3N2	1.312
T3S1L1N1	T3	0.37	0.401	T3S1L1N2	0.398
T3S1L2N1	T3	0.60	0.679	T3S1L2N2	0.678
T3S1L3N1	T3	1.29	1.270	T3S1L3N2	1.317
T3S2L1N1	T3	0.37	0.394	T3S2L1N2	0.413
T3S2L2N1	T3	0.60	0.688	T3S2L2N2	0.681
T3S2L3N1	T3	1.29	1.258	T3S2L3N2	1.280

* Specimen ID = fire condition (T0–T3) + smoke condition (S0–S2) + target DFT (L1–L3) + sample numbe r(N1–N2).

three large space fire curves (T1–T3); S0 to S2 stand for the three different smoke conditions, by using 0, 10 and 20 kg/m^2 of oak per floor area of the furnace respectively; L1 to L3 represent the different target DFTs; and N1 and N2 indicate duplicate specimens. The average measured DFTs of intumescent coatings for the test specimens are also included in Table 3.2 (Han 2017, 97–98).

3.2.2 Fire protection

A single-component solvent-based intumescent coating was used for all the test specimens. The binder of the coatings was acrylic. The target DFTs were 0.37, 0.60 and 1.29 mm, which are typical in practical applications to achieve fire resistance ratings of 0.5, 1.0 and 1.5 h separately. The coatings were applied on the fire exposure side of the steel plates. The applied intumescent coatings on the steel plate specimens were dried at ambient temperature for at least 48 h before fire testing. The actual DFTs of each specimen were measured at ten locations by a digital thickness gauge and the differences between all the measured DFT values and the average values were less than 5%.

Two types of materials (mineral wool at least 150 mm thick and vicuclad 50 mm thick) were attached to the steel plates to minimize heat loss on the backside of the samples, as shown in Figure 3.3. Although the mineral wool was able to provide the required insulation, vicuclad boards were used to help fix the mineral wool to the steel plates for better insulation.

3.2.3 Test setup

The tests were carried out in two identical screen furnaces as shown in Figure 3.4 (Han et al. 2019, 5). The furnaces measured 2m × 0.5m × 0.5m in internal dimensions, and were heated by a gas burner. The interior surfaces of the furnaces were lined with 200-mm-thick ceramic fibre materials to efficiently transfer heat to the specimens. There was a window on one side of the furnace to allow the test specimens to be observed during the test. Two specimens were tested each time in one furnace. They were attached to the furnace wall or ceiling so that they could be considered to be exposed to fire on one side only. Owing to the small volume of the furnaces, uniformity of heating was easily achieved, ensuring that the samples on the walls and ceiling of the furnace received very similar heat flux. Pre-tests on unprotected steel plates showed that those on the ceiling and on the wall attained very similar temperatures, which confirmed the homogeneity of heating inside the furnace. The furnace was controlled in such a way that the average temperature of the two thermocouples near the furnace roof followed the theoretical temperature–time curve. Because the furnaces were used for research purposes only, and the main temperature of interest was the steel temperature, plate thermometers were not used.

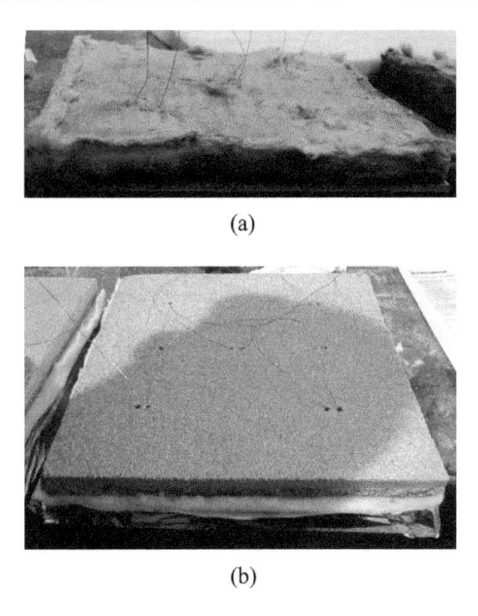

Figure 3.3 Fire protection of test specimens. (a) Mineral wool to minimize heat loss. (b) Vicuclad used above the mineral wool.

Figure 3.4 Screen furnaces. (a) Inside. (b) Outside.

3.3 EXPERIMENTAL MEASUREMENTS AND OBSERVATIONS OF INTUMESCENT COATING BEHAVIOUR

The fire tests were terminated when the steel temperature reached 600, 450, 400 and 325°C for tests T0 to T4 respectively. Typical surface conditions of the intumescent coatings tested under different fire conditions are shown in Figure 3.5. Consistently, the intumescent coatings tested under the ISO standard fire condition fully expanded and oxidized to form a white char. The char was continuous and smooth and the bubbles appeared to be small and evenly distributed. Under large space fire condition T1, since the fire and steel temperatures were still reasonably high, the expanded intumescent coatings were similar with full expansion under the ISO standard fire

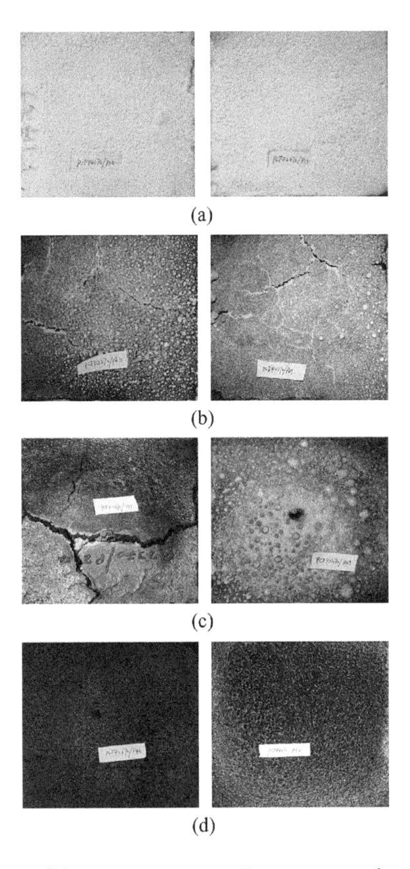

Figure 3.5 Appearance of intumescent coatings exposed to different fires after test. (a) ISO standard fire test (left T0S1L2N2, right T0S2L2N1). (b) Large space fire T1 test (left T1S1L2N1, right T1S2L2N1). (c) Large space fire T2 test (left T2S1L2N1, right T2S1L3N1). (d) Large space fire T3 test (left T3S1L2N2, right T3S2L2N2).

condition. However, cracks appeared on the char surface and the bubbles were different in size. The intumescent coatings tested under large space fire condition T2 behaved progressively worse, as indicated by the large cracks and large bubbles. As for the case of large space fire condition T3, the intumescent coatings did not expand fully, since the fire and steel temperatures were very low. These observations clearly indicate a need to distinguish large space fires from the ISO 834 fire when assessing intumescent coating behaviour.

Intumescent coating expansions were carried out at ten locations without large bubbles or cracks. For the fire condition T0 and T1 tests with relatively high temperatures, the intumescent coating expansions of the test specimens were relatively uniform and the measured expansions were within ±10% of the average, as shown in Figure 3.5(a, b). For the fire condition tests T2 and T3 with relatively low temperatures, some large bubbles appeared on the surfaces of the specimens and the intumescent coating expansions were not so uniform, as shown in Figure 3.5(c, d). Table 3.3 lists the average post-fire thicknesses and expansion ratios of the intumescent coatings. Obviously, the intumescent coating expansion decreases as the maximum fire temperature decreases (from T0 to T3).

Table 3.3 Expanded thicknesses and expansion ratios of test specimens

Fire condition	*Target DFT (mm)*	*Average measured DFT (mm)*	*Average expanded thickness (mm)*	*Expansion ratio*	*Average Expansion ratio*
T0	0.370	0.395	8	20.25	
	0.600	0.664	15	22.59	21.62
	1.290	1.317	29	22.02	
	0.370	0.403	8.3	20.60	
	0.600	0.671	12.5	18.63	20.02
	1.290	1.320	27.5	20.83	
	0.370	0.411	7.5	18.25	
	0.600	0.668	12	17.96	19.25
	1.290	1.300	28	21.54	
T1	0.370	0.403	7	17.37	
	0.600	0.678	8	11.80	13.80
	1.290	1.308	16	12.23	
	0.370	0.399	7	17.54	
	0.600	0.678	8.5	12.54	12.99
	1.290	1.293	11.5	8.89	
	0.370	0.403	7	17.37	
	0.600	0.684	8	11.70	14.55
	1.290	1.269	18.5	14.58	

(Continued)

Table 3.3 (Continued) Expanded thicknesses and expansion ratios of test specimens

Fire condition	*Target DFT (mm)*	*Average measured DFT (mm)*	*Average expanded thickness (mm)*	*Expansion ratio*	*Average Expansion ratio*
T2	0.370	0.407	3	7.37	
	0.600	0.607	6	9.88	9.14
	1.290	1.278	13	10.17	
	0.370	0.399	3.5	8.77	
	0.600	0.678	6.5	9.59	9.13
	1.290	1.328	12	9.04	
	0.370	–	–	–	
	0.600	0.681	7	10.28	9.03
	1.290	1.286	10	7.78	
T3	0.370	0.401	–	–	
	0.600	0.650	–	–	–
	1.290	1.319	–	–	
	0.370	0.403	–	–	
	0.600	0.684	–	–	–
	1.290	1.296	–	–	
	0.370	0.405	–	–	
	0.600	0.660	–	–	–
	1.290	1.317	–	–	

3.4 THE THREE-STAGE MODEL OF THERMAL CONDUCTIVITY

3.4.1 Temperature dependent effective thermal conductivity

The temperature dependent effective thermal conductivities of the intumescent coatings for the test specimens were obtained using Equation (2.5). Figure 3.6(a–d) compares the effective thermal conductivity–temperature relationships obtained from the tests under different fire conditions. It can be seen that the relationships of the intumescent coatings are similar for different coating thicknesses. Therefore, an average thermal conductivity–temperature relationship can be used for each fire condition test.

Figure 3.7 compares the average thermal conductivity–temperature relationships of the intumescent coatings under different fire conditions. Generally, the thermal conductivities of the coatings reduce with decrease in the maximum fire temperature. It can be seen that when the steel temperature is below 300°C, the thermal conductivities of the coatings under different fire conditions are varied and differentiated substantially. However, at higher steel temperatures over 300°C, the variations of thermal conductivities under different fire conditions are not substantial and appear similar.

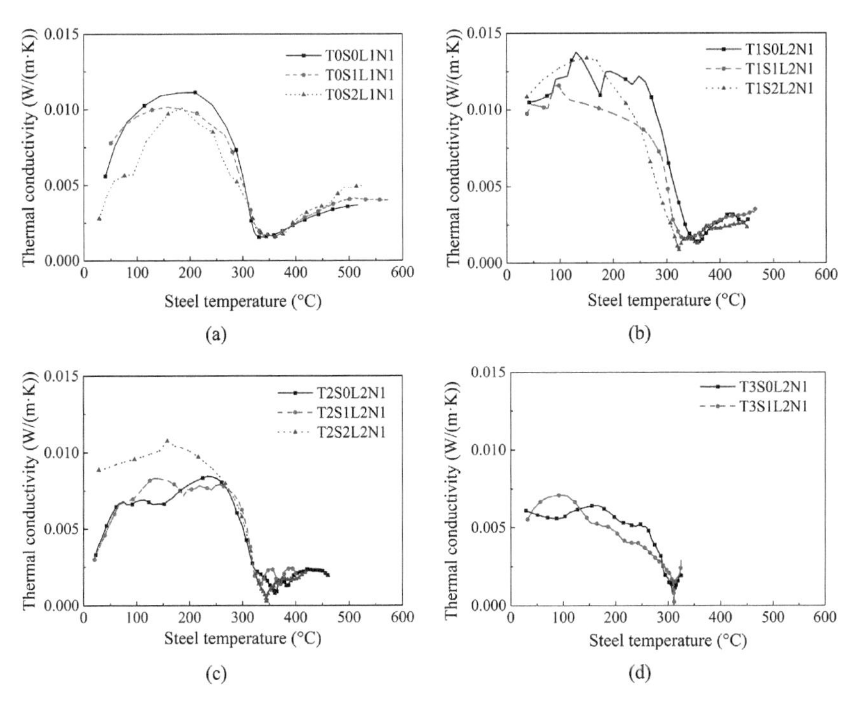

Figure 3.6 Effective thermal conductivities of intumescent coatings under different fire conditions. (a) Standard fire T0. (b) Large space fire T1. (c) Large space fire T2. (d) Large space fire T3.

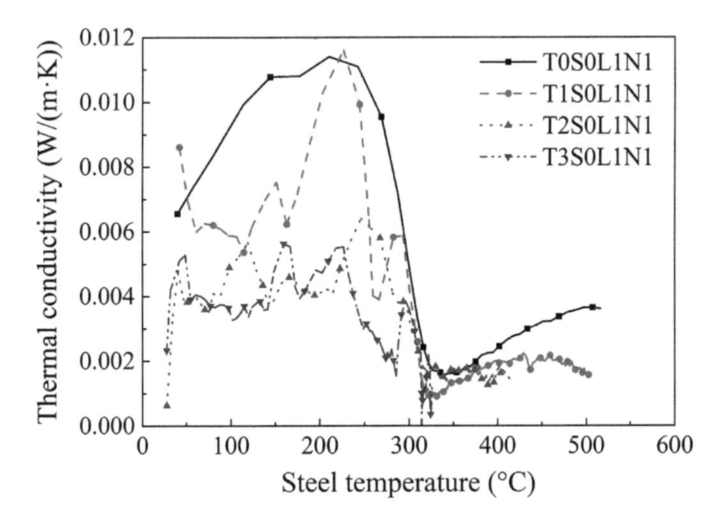

Figure 3.7 Comparison between average effective thermal conductivity–temperature relationships under different fire conditions.

3.4.2 Constant effective thermal conductivities

Although the thermal conductivities of the intumescent coatings shown in Figures 3.6 and 3.7 gave faithful values based on the fire tests, they are temperature dependent and inconvenient for use in temperature calculations of protected steel elements exposed to fire. Therefore, a simplified thermal conductivity model is necessary for intumescent coatings subjected to large space fires.

It is observed that the performance of intumescent coatings can be divided into a melting stage, an expanding stage and a fully expanded stage with increasing temperature of the protected steel substrate, as shown in Figure 3.8 (Griffin 2010, 253). At about 100°C, intumescent coatings degrade to form a viscous liquid. At such a temperature, the intumescent coatings have not started expansion so their thermal conductivity is high, as shown in Figure 3.9. When the intumescent coating temperature reaches

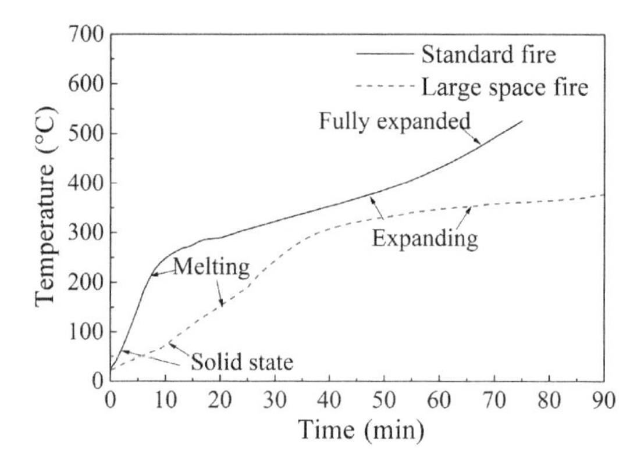

Figure 3.8 Three-stage performance of intumescent coatings with protected steel temperature elevation.

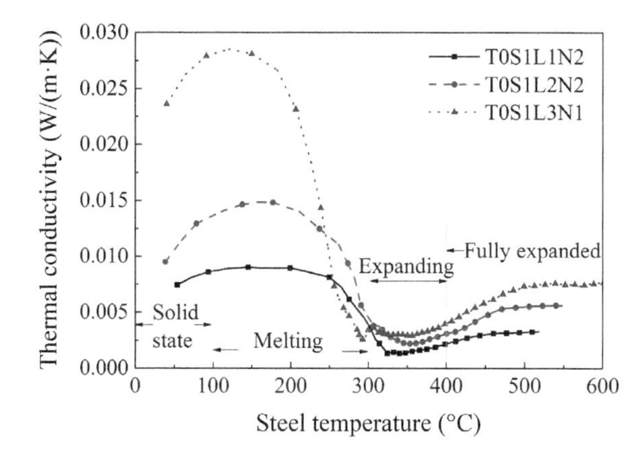

Figure 3.9 Typical temperature dependent thermal conductivity of coatings with various dry film thicknesses.

about 300°C, the blowing agent within the intumescent coatings decomposes and large amounts of gas are released within the molten matrix to cause rapid expansion of the coatings. This is shown by the drastic reduction in intumescent coatings' effective thermal conductivity at steel temperatures above 300°C. When the temperature is over 400°C, the coatings are fully expanded so their effective thermal conductivity stabilizes and gradually increases due to void radiation.

According to intumescent coating behaviour in fire, the temperature dependent thermal conductivities of such coatings may be divided into three stages: melting (when the temperature is less than 300°C), expanding (when less than 400°C) and full expansion (when more than 400°C). Based on the above observation, a three-stage constant effective thermal conductivity model is proposed for intumescent coatings exposed to large space fires.

The constant effective thermal conductivities of intumescent coatings at various stages are defined as follows.

Fully expanded stage:

$$\lambda_e = \frac{1}{200} \cdot \int_{T_s=400}^{T_s=600} \lambda(T_s)\,dT_s \tag{3.1}$$

Expanding stage:

$$\lambda_{\mathrm{e-b}} = \frac{1}{100} \cdot \int_{T_s=300}^{T_s=400} \lambda(T_s)\,dT_s \tag{3.2}$$

Melting stage:

$$\lambda_{\mathrm{e-m}} = \frac{1}{200} \cdot \int_{T_s=100}^{T_s=300} \lambda(T_s)\,dT_s \tag{3.3}$$

where λ_e, $\lambda_{\mathrm{e-b}}$, $\lambda_{\mathrm{e-m}}$ are the constant effective thermal conductivity (W $m^{-1} \cdot K^{-1}$) at the fully expanded stage, expanding stage and melting stage respectively; and $\lambda(T_s)$ is the temperature dependent effective thermal conductivity (W $m^{-1} \cdot K^{-1}$).

The three-stage constant effective thermal conductivity model is illustrated in Figure 3.10. Using the temperature of the steel substrate instead of the intumescent coatings for quantifying the constant effective thermal conductivities of the latter is a conservative consideration for practical application, because the temperature of the steel substrate will not be greater than that of the intumescent coatings, and it is the temperature of the steel substrate that is of interest for the fire safety of steel structures.

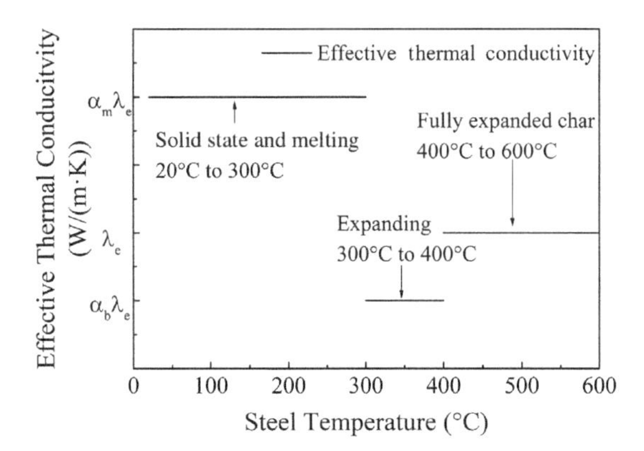

Figure 3.10 Three-stage constant effective thermal conductivity model for large space fires.

3.4.3 Inter-fire relationships for three-stage constant effective thermal conductivities

The results of the constant effective thermal conductivities for the intumescent coatings at three different stages for all the test specimens are listed in Table 3.4. The values at the fully expanded stage for fire conditions T2 and T3 are missing because the intumescent coatings were not fully expanded. Figure 3.11 compares the average values of the constant effective thermal conductivities for the specimens with a 0.6 mm DFT of intumescent coating under different fire conditions. It is seen that these conductivities of the intumescent coatings at the expanding and fully expanded stages are similar under various fire conditions, while the values at the melting stage are much different from the standard fire exposure to various large space fire conditions. Therefore, the constant effective thermal conductivity of the intumescent coatings at the fully expanded stage obtained under the ISO standard fire (fire condition T0) can be proposed for use in any other fire conditions when the protected steel temperature is over 400°C, with fully expanded coatings.

In order to obtain the constant effective thermal conductivities of intumescent coatings at the melting and expanding stages, the following inter-stage relationship is established on the basis of the fully expanded stage as

$$\lambda_e(T_s) = \begin{cases} \lambda_{e-m} = \alpha_m \lambda_e & T_s \leq 300^\circ\text{C} \\ \lambda_{e-b} = \alpha_b \lambda_e & 300^\circ\text{C} < T_s < 400^\circ\text{C} \\ \lambda_e & T_s \geq 400^\circ\text{C} \end{cases} \tag{3.4}$$

where λ_{e-m}, λ_{e-b}, λ_e are the constant effective thermal conductivities of intumescent coatings at the melting stage, expanding stage and full expansion stage respectively; and α_m and α_b are the ratios of the constant effective thermal conductivities of intumescent coatings at the melting stage and expanding stage to that at the full expansion stage respectively.

Table 3.4 Constant effective thermal conductivities of intumescent coatings at three stages

Specimen ID	*Conductivity at melting stage (W/m·K)*	*Conductivity at expanding stage (W/m·K)*	*Conductivity at fully expanded stage (W/m·K)*
T0S0L1N1	0.0099	0.0017	0.0032
T0S0L1N2	0.0100	0.0019	0.0034
T0S1L1N1	0.0094	0.0022	0.0036
T0S1L1N2	0.0086	0.0018	0.0030
T0S2L1N1	0.0078	0.0021	0.0039
T0S2L1N2	0.0084	0.0018	0.0032
T0S0L2N1	—	—	—
T0S0L2N2	—	—	—
T0S1L2N1	0.0126	0.0026	0.0039
T0S1L2N2	0.0136	0.0027	0.0048
T0S2L2N1	0.0159	0.0025	0.0046
T0S2L2N2	0.0154	0.0021	0.0040
T0S0L3N1	0.0233	0.0032	0.0066
T0S0L3N2	0.0237	0.0034	0.0068
T0S1L3N1	0.0229	0.0032	0.0065
T0S1L3N2	0.0190	0.0032	0.0062
T0S2L3N1	0.0168	0.0029	0.0060
T0S2L3N2	0.0199	0.0031	0.0056
T1S0L1N1	0.0071	0.0014	0.0019
T1S0L1N2	0.0077	0.0015	0.0020
T1S1L1N1	0.0063	0.0019	0.0021
T1S1L1N2	0.0058	0.0022	0.0022
T1S2L1N1	0.0074	0.0019	0.0026
T1S2L1N2	0.0072	0.0019	0.0025
T1S0L2N1	0.0117	0.0024	0.0029
T1S0L2N2	0.0116	0.0022	0.0027
T1S1L2N1	0.0119	0.0020	–
T1S1L2N2	0.0125	0.0019	–
T1S2L2N1	0.0125	0.0017	0.0025
T1S2L2N2	0.0121	0.0017	–
T1S0L3N1	0.0157	0.0032	0.0040
T1S0L3N2	0.0194	0.0025	–
T1S1L3N1	0.0153	0.0029	0.0041
T1S1L3N2	0.0132	0.0030	0.0037
T1S2L3N1	0.0202	0.0025	0.0038
T1S2L3N2	0.0199	0.0023	–
T2S0L1N1	0.0044	0.0017	
T2S0L1N2	0.0050	0.0018	
T2S1L1N1	0.0059	0.0023	
T2S1L1N2	0.0085	0.0023	

(Continued)

Table 3.4 (Continued) Constant effective thermal conductivities of intumescent coatings at three stages

Specimen ID	*Conductivity at melting stage (W/m·K)*	*Conductivity at expanding stage (W/m·K)*	*Conductivity at fully expanded stage (W/m·K)*
T2S2L1N1			
T2S2L1N2			
T2S0L2N1	0.0070	0.0016	
T2S0L2N2	0.0077	0.0017	
T2S1L2N1	0.0073	0.0021	
T2S1L2N2	0.0108	0.0025	
T2S2L2N1	0.0106	0.0013	
T2S2L2N2	0.0098	0.0011	
T2S0L3N1	—	—	
T2S0L3N2	—	—	
T2S1L3N1	0.0132	0.0026	
T2S1L3N2	0.0193	0.0035	
T2S2L3N1	0.0151	0.0021	
T2S2L3N2	0.0183	0.0027	
T3S0L1N1	0.0041	0.0014	
T3S0L1N2	0.0049	0.0012	
T3S1L1N1	0.0031	0.0007	
T3S1L1N2	0.0059	0.0013	
T3S2L1N1	0.0078	0.0011	
T3S2L1N2	0.0073	0.0012	
T3S0L2N1	0.0057	0.0013	
T3S0L2N2	0.0068	0.0014	
T3S1L2N1	0.0053	0.0012	
T3S1L2N2	0.0091	0.0018	
T3S2L2N1	0.0156	0.0027	
T3S2L2N2	0.0194	0.0031	
T3S0L3N1	0.0111	0.0022	
T3S0L3N2	0.0130	0.0026	
T3S1L3N1	0.0112	0.0020	
T3S1L3N2	0.0170	0.0022	
T3S2L3N1	0.0113	0.0043	
T3S2L3N2	0.0161	0.0052	

The values of the factors α_m and α_b for the different fire conditions can be obtained from the test measurement, as shown in Table 3.5. The results suggest that the value of α_b, the ratio of the constant effective thermal conductivity at the expanding stage to that at the fully expanded stage, is roughly 0.5. This is reasonable because when the steel temperature is in the range 300–400°C, the expanding of the intumescent coatings consumes a

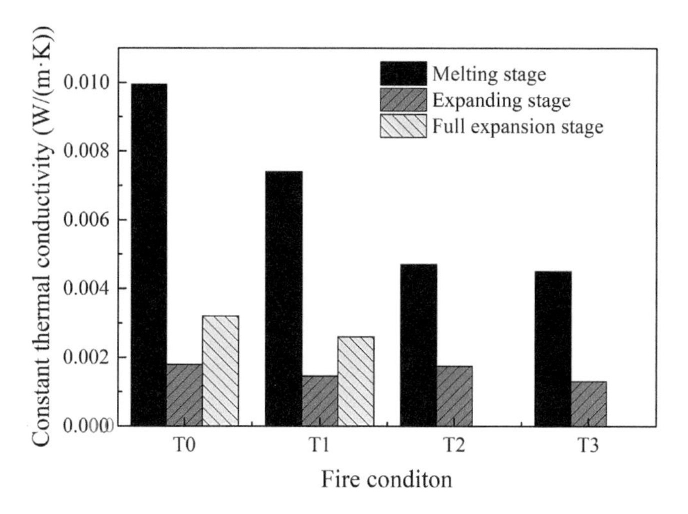

Figure 3.11 Comparison of average constant effective thermal conductivities of intumescent coatings under various fire conditions.

Table 3.5 Ratios of constant effective thermal conductivities for different fire conditions

Fire condition	*Melting stage*	*Expanding stage*	*Full expansion stage*
ISO standard fire T0	3.26	0.54	1.00 (based on fire T0)
Large space fire T1	2.69	0.48	
Large space fire T2	2.46	0.47	
Large space fire T3	1.75	0.43	

large amount of heat due to chemical reaction to reduce the thermal conductivity, as shown in Figure 3.6.

The values of α_m are greater than 1, as shown in Table 3.5, because the intumescent coatings at the melting stage are not fully effective at resisting heat transfer with the very thin dry film. It can be seen that the values of α_m vary with different fire conditions, but follow a trend of decreasing as the maximum fire gas temperature decreases (from T0 to T3).

3.5 APPLICABILITY OF THE THREE-STAGE CONSTANT EFFECTIVE THERMAL CONDUCTIVITY MODEL

3.5.1 Calculation of protected steel temperature with the three-stage thermal model

The constant effective thermal conductivities of intumescent coatings vary at different stages, as illustrated in Figure 3.10, where the temperature calculation of the protected steel exposed to various fire conditions can be greatly simplified.

The protected steel temperature can be calculated by:

$$T_s(t+\Delta t)=T_s(t)+\Delta T_s \tag{3.5}$$

ΔT_s can be calculated by the following equations:

$$\Delta T_s=\frac{\left(T_g(t+\Delta t)-T_s(t)\right)A_i / V}{\left(d_i / \lambda_{e-m}\right)c_s\rho_s}\Delta t \quad T_s \leq 300^{\circ}\mathrm{C} \tag{3.6}$$

$$\Delta T_s=\frac{\left(T_g(t+\Delta t)-T_s(t)\right)A_i / V}{\left(d_i / \lambda_{e-b}\right)c_s\rho_s}\Delta t \qquad 300^{\circ}\mathrm{C}<T_s<400^{\circ}\mathrm{C} \tag{3.7}$$

$$\Delta T_s=\frac{\left(T_g(t+\Delta t)-T_s(t)\right)A_i / V}{\left(d_i / \lambda_{e}\right)c_s\rho_s}\Delta t \qquad T_s \geq 400^{\circ}\mathrm{C} \tag{3.8}$$

where $T_s(t)$ is the steel temperature at time t; $T_s(t+\Delta t)$ is the steel temperature at time $t+\Delta t$; $T_g(t+\Delta t)$ is the gas temperature at time $t+\Delta t$; Δt is the time interval for every step, which should be less than 30s; λ_{e-m}, λ_{e-b}, λ_e are the constant effective thermal conductivities of intumescent coatings at the melting stage, expanding stage and full expansion stage respectively.

3.5.2 Effectiveness of the three-stage thermal conductivity model

To verify the effectiveness of this model for predicting the temperature of the protected steel substrates under various fire conditions, as described in Section 3.4, the steel temperature of the test specimens presented in Section 3.2 is calculated and compared with those measured from the tests. Figure 3.12 compares the calculated steel temperatures using the constant effective thermal conductivity values given in Table 3.5 with the test results under different fire conditions. It can be seen that in all cases, steel temperatures in the expanding and expanded stages (>300°C) respectively, obtained from calculation and test measurement, closely match each other. At the melting stage when the steel temperature is less than 300°C, the calculated steel temperatures are still in good agreement with the test results for all the large space fire conditions (T1–T3), though the difference is significant for the standard fire condition (T0). However, the interest in steel temperatures for structural safety under the standard fire is not at the melting stage of the coating protection, but at the fully expanded stage.

The above results demonstrate that it is feasible to employ the three-stage constant effective thermal conductivity model to quantify the heat insulation performance of intumescent coatings under large space fires. However, it should be noted that what is proposed is product dependent and the process

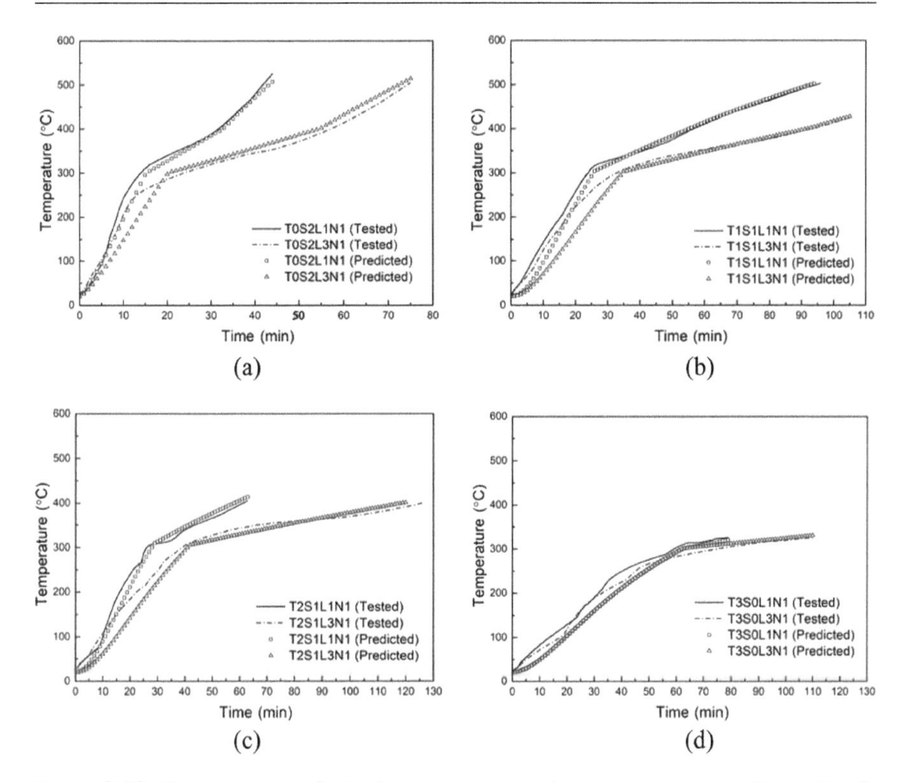

Figure 3.12 Comparison of steel temperatures between test results and calculated using three-stage constant effective thermal conductivity model. (a) Standard Fire T0. (b) Large space fire T1. (c) Large space fire T2. (d) Large space fire T3.

described in this book would need to be conducted on other intumescent coating products. Furthermore, the conductivity is a simple quantity for the convenience of fire engineering application in the safety evaluation of steel structures protected with intumescent coatings.

3.6 SUMMARY

The main findings of this chapter can be summarized as follows.

1. Intumescent coatings have different thermal resistance performances under large space fires. In general, under a large space fire in particular, the performance of these coatings can be divided into three stages: melting, expanding and fully expanded. The melting stage may be considered to have completed and the expanding stage to start when the protected steel temperature has reached 300°C; the fully expanded stage may be considered to have been formed when the protected steel temperature has reached 400°C.

2. The effective thermal conductivity of intumescent coatings at all the stages of melting, expanding and full expansion is temperature dependent. The variation of the effective thermal conductivity during the melting stage is much more significant than that during the expanding and full expansion stages.
3. A three-stage constant effective thermal conductivity model has been proposed to simply quantify the thermal resistance property of intumescent coatings. A constant thermal conductivity value may be defined as the mean value of the temperature-dependent effective thermal conductivity in the steel substrate temperature range of 100–300°C for the melting stage, 300–400°C for the expanding stage, and 400–600°C for the fully expanded stage.
4. The constant effective thermal conductivities of intumescent coatings at the fully expanded stage for different fire conditions are similar. The ratio of the conductivity at the expanding stage to that at the fully expanded stage is around 0.5. The ratio of the conductivity at the melting stage to that at the fully expanded stage is greater than 1.0 and increases with the increase in the maximum fire gas temperature.
5. Applicability of the three-stage constant effective thermal conductivity model has been verified for predicting the steel substrate temperatures of test specimens protected with an intumescent coating under various large space fire conditions.

REFERENCES

CECS. 2006. *CECS 200:2006.Technical Code for Fire Safety of Steel Structure in Buildings*. Beijing: China Planning Press.

Du, Yong. 2007. *A Practical Approach for Fire Resistance Design of Large Space Building Grid Structures*. Shanghai: Tongji University.

Griffin, G. J. 2010. “The Modeling of Heat Transfer across Intumescent Polymer Coatings.” *Journal of Fire Sciences* 28(3): 249–277.

Han, Jun. 2017. *Study on the Performance of Intumescent Coating under Large Space Fires*. Shanghai: Tongji Unviersity

Han, Jun, Li, Guo-Qiang, Wang, Yong C., and Xu, Qing. 2019. “An Experimental Study to Assess the Feasibility of a Three Stage Thermal Conductivity Model for Intumescent Coatings in Large Space Fires.” *Fire Safety Journal* 109: 102860. https://doi.org/10.1016/j.firesaf.2019.102860.

Li, Guo-Qiang, Xu, Qing, Han, Jun, Zhang, Chao, Lou, Guo-Biao, Jiang, Shou-Chao, and Jiang, Jian. 2018. “Fire Safety of Large Space Steel Buildings.” *International Conference on Engineering Research and Practice for Steel Construction 2018 (ICSC2018)*, Hong Kong, China, 5 to 7 September.

Li, Guoqiang, and Wang, Peijun. 2013. “Fire in Buildings.” In *Advanced Analysis and Design for Fire Safety of Steel Structures*, edited by Guoqiang Li and Peijun Wang, 11–36. Berlin, Heidelberg: Springer Berlin Heidelberg.

Lucherini, Andrea, Giuliani, Luisa, and Jomaas, Grunde. 2018. "Experimental Study of the Performance of Intumescent Coatings Exposed to Standard and Non-Standard Fire Conditions." *Fire Safety Journal* 95: 42–50. https://doi.org/10.1016/j.firesaf.2017.10.004.

Wang, Lingling, Dong, Yuli, Zhang, Chao, and Zhang, Dashan. 2015. "Experimental Study of Heat Transfer in Intumescent Coatings Exposed to Non-Standard Furnace Curves." *Fire Technology* 51(3): 627–643. https://doi.org/10.1007/s10694-015-0460-7.

Yuan, Ji-Feng. 2009. *Intumescent Coating Performance on Steel Structures under Realistic Fire Conditions*. Manchester: University of Manchester.

Chapter 4

Behaviour of intumescent coatings exposed to localized fires

Different from the approximately uniform distribution of the hot smoke temperature released from fire in a large space building, the gas temperature of a localized fire apart from the flame is substantially non-uniform (Zhang and Li 2012, 125; Byström et al. 2014, 148). So, understanding the behaviour of intumescent coatings exposed to localized fire is also necessary.

This chapter presents the experimental results from a programme of fire tests to investigate the behaviour of intumescent coatings applied to steel hollow section members under localized fire exposures. Two types of intumescent coatings, waterborne and solvent-based, were employed on the steel member specimens. The behaviour of the coatings exposed to localized fire is compared with that under the ISO standard fire and large space fires (Xu 2021, 47–70).

4.1 LOCALIZED FIRE AND TEST SETUP

The localized fire tests were conducted on two types of steel member specimens protected with intumescent coatings: one type was a horizontal member specimen representing steel beams, and the other type was a vertical member specimen representing steel columns (Xu et al. 2020, 2). The beam specimen was simply supported on two steel stubs and the pool fire was located beneath the mid-span of the specimen at one metre below its lower flange, as shown in Figure 4.1 (Xu et al. 2020, 4). A couple of column specimens were fixed to the ground and placed by the two sides of the fire pool, as shown in Figure 4.2 (Xu et al. 2020, 5).

Pool fires were used in the tests to simulate localized fires. The metal box for pool fires was square, measuring 1 m by 1 m in plan and 0.2 m in depth, equivalent to a circular fire source with a diameter of 1.13 m in terms of the surface area. N-heptane was used as fuel because of its high calorific value and low smoke density.

DOI: 10.1201/9781003287919-4

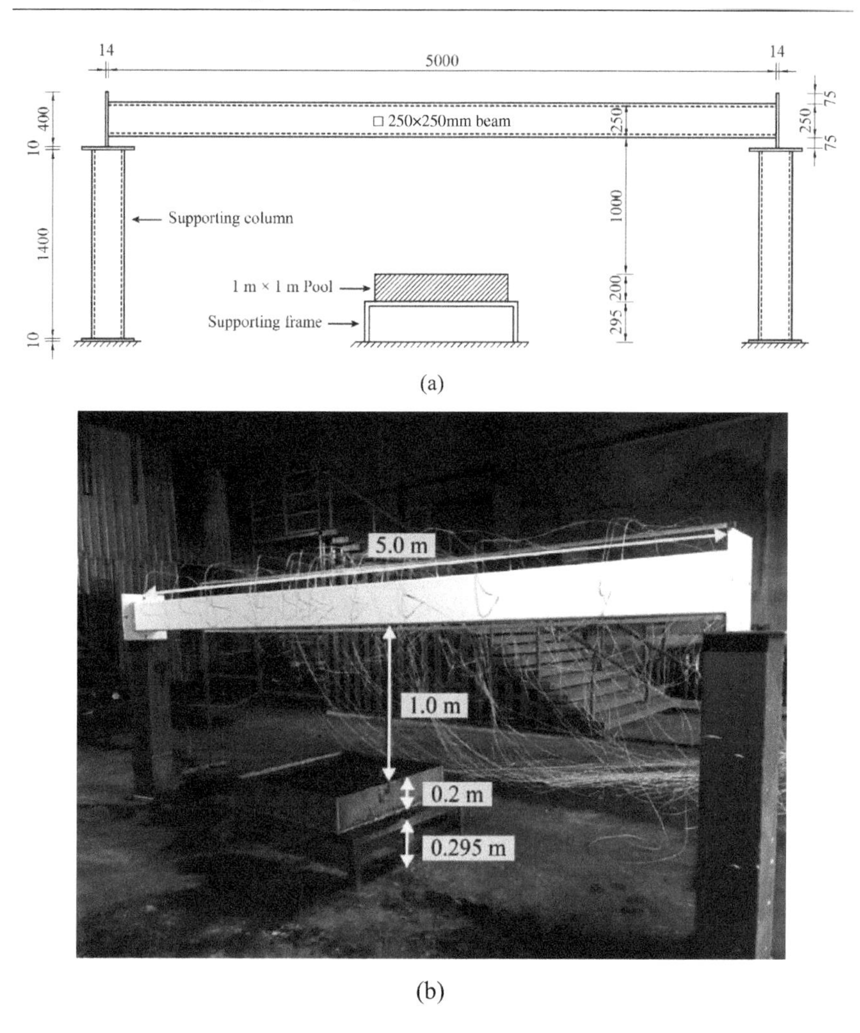

Figure 4.1 Test setup for horizontal member (beam) specimen (unidentified units: mm). (a) Test setup (units: mm). (b) Test specimen before fire testing.

A localized fire was designed to achieve a heat release rate (HRR) of approximately 2 MW and a steady burning duration of about 30 minutes. The HRR can be related to the mass loss rate according to Equation (4.1):

$$\dot{Q}(t) = \dot{m}(t)\chi_{\text{eff}}\Delta H_c \tag{4.1}$$

where ΔH_c is the heat of combustion, $\dot{m}(t)$ is the mass loss rate, and χ_{eff} is the combustion efficiency, assumed to be 0.8 (Hostikka et al. 2001).

Since the heat of combustion of heptane is 44.6 MJ/kg (Heskestad 2016, 396–428), the estimated mass loss rate, $\dot{m}(t)$, to achieve an HRR of 2 MW

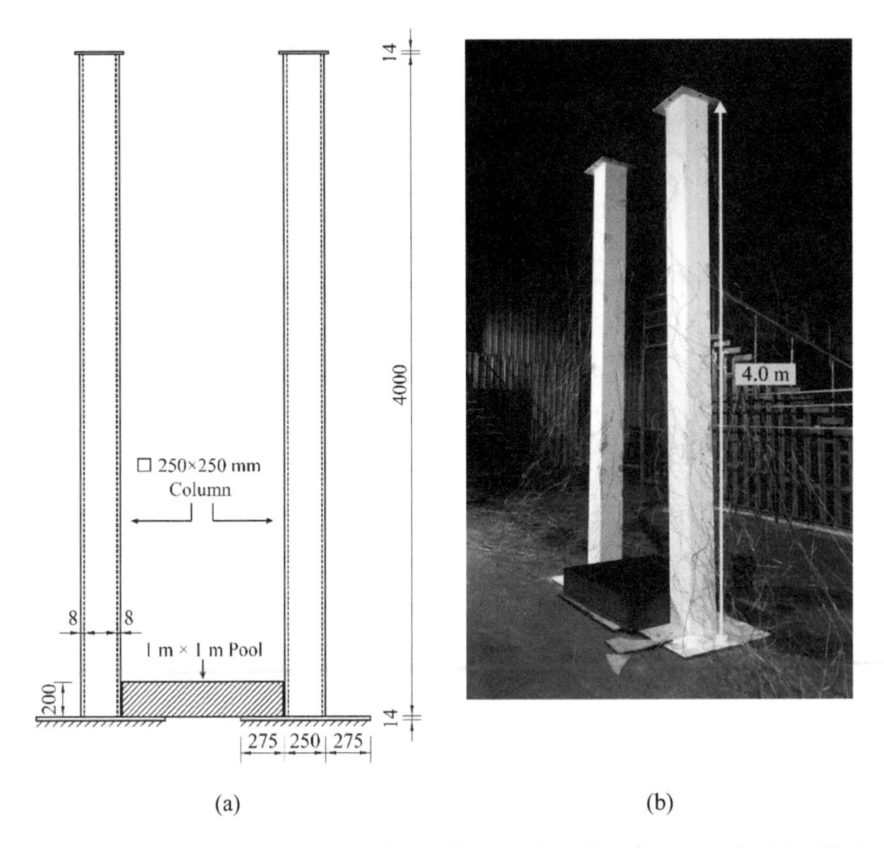

Figure 4.2 Test setup for vertical member (column) specimens (unidentified units: mm). (a) Test setup (b) Specimens before test.

Table 4.1 Information of localized fires for different test specimens

Test no.	*Specimen no.*	*Specimen type*	*Fuel volume (L)*	*Duration (min)*	*Estimated HRR (MW)*	*Measured average HRR (MW)*
1	BW	Horizontal	140	27.5	2.57	2.07
2	BS	Horizontal	140	27.5	2.57	2.07
3	CS1, CW2	Vertical	130	24.2	2.57	2.18
4	CW1, CS2	Vertical	135	25.5	2.57	2.15

is approximately 0.072 kg/s , which can be provided by an N-heptane pool fire with a diameter of 1.13 m.

Table 4.1 (Xu et al. 2020, 4) provides details of the measured burning time, and the estimated and average measured heat release rates of the localized fires for all the test specimens.

4.2 TEST SPECIMENS

4.2.1 Steel members

Square hollow steel tubes with a section of 250 × 250 × 8 mm were used for both beam and column specimens for the localized fire tests (Xu et al. 2020, 2). The section factor of the steel members used was 132.98 m^{-1}. Prior studies have shown that the shape of steel member sections has little influence on the performance of intumescent coatings (de Silva et al. 2019, 232–244). The two beam specimens were each 5 m in length, and the four column specimens were 4 m in height. The specimens were unloaded and welded end plates added for stability during testing.

4.2.2 Thermocouples

The steel temperatures of the beam and column specimens and surrounding gas temperatures were measured by Type-K inconel sheathing thermocouples with a diameter of 2 mm at selected locations. Figure 4.3 (Xu et al. 2020, 2) shows the locations of thermocouples installed at 11 cross-sections

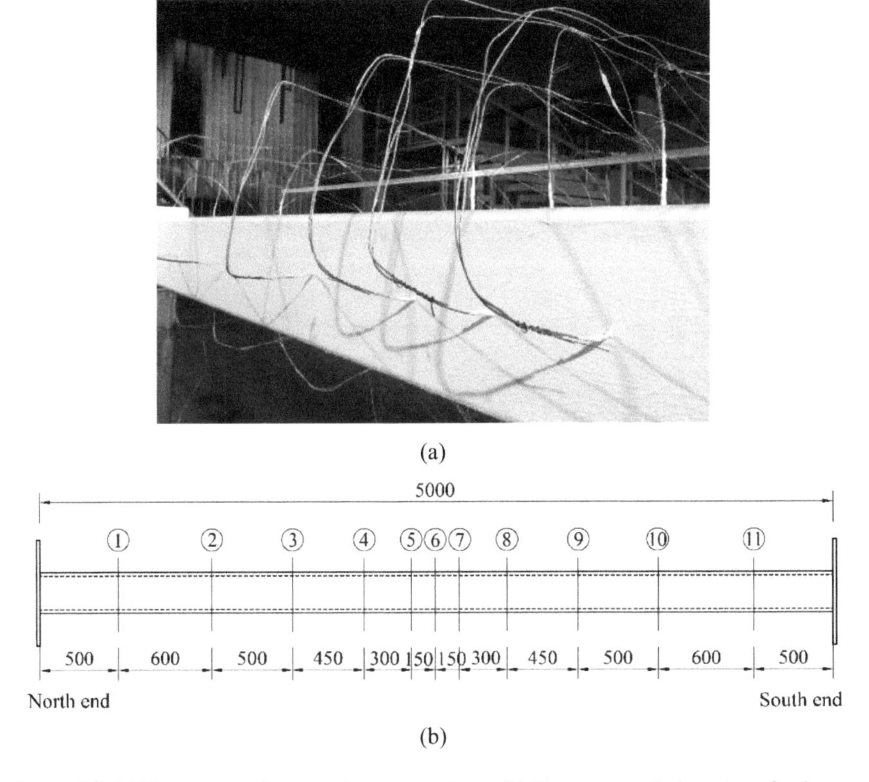

Figure 4.3 (a) Thermocouples on a beam specimen. (b) Thermocouple locations for beam specimens (units: mm).

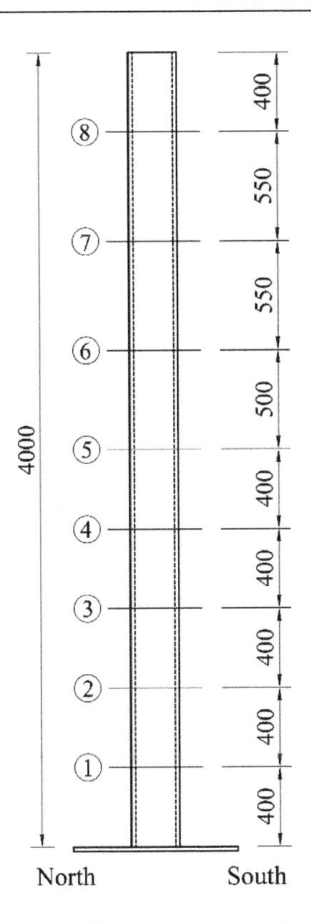

Figure 4.4 Thermocouple stations for column specimens (units: mm).

along the beam specimens, and Figure 4.4 (Xu et al. 2020, 3) shows the eight cross-sections at the stations for the column specimens. These cross-sections of thermocouples allowed the detailed measurement of temperatures under the experimental heating conditions. Due to large temperature gradients close to the fire source, the thermocouple stations near the centres of the beam specimens were more closely spaced.

At each cross-section of both beam and column specimens, eight thermocouples were installed, as shown in Figure 4.5 (Xu et al. 2020, 3), four of which were embedded in the steel member to measure steel temperatures. This method of thermocouple installation was chosen to ensure that the thermocouples stayed in place and to avoid any disturbance to intumescent coatings. The other four thermocouples at each cross-section of the steel members were located approximately 70 mm from the steel member surface to measure the surrounding gas temperatures. The thermocouples measuring steel temperatures were denoted "TS-section number-surface name" and those measuring gas temperatures were "TG- section number- surface name". The front surface of the member facing the fire was named "F", the back surface named "B" and the two side surfaces were named "S1" and "S2".

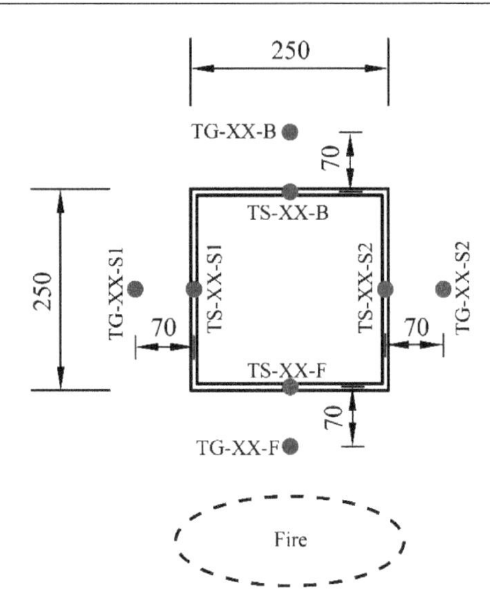

Figure 4.5 Locations of thermocouples at each cross-section of steel member specimens.

Note: TG-Thermocouples for gas temperature;TS-Thermocouples for steel temperature (units: mm).

4.2.3 Fire protection

Two types of intumescent coatings were applied to the steel member specimens, including a single component waterborne intumescent coating and a single component solvent-based acrylic intumescent coating. The target dry film thickness (DFT) for both types of intumescent coatings was 0.6 mm on all the specimens. This DFT is typical of that required to achieve a standard fire resistance rating of 30 min for the coatings used for a steel limiting temperature of 500°C, which is also the designed duration of the pool fire for the localized fire tests. The actual installed DFT of the intumescent coatings was measured at 150 locations for each specimen, and the average value was taken as the measured DFT for each member specimen. Table 4.2

Table 4.2 Summary of the measured DFTs for the test specimens

Specimen no.	*Specimen type*	*Length (m)*	*Coating type*	*Number of cross-sections for temperature measurement*	*Average coating DFT/(μm)*	*Standard deviation of DFT/ (μm)*
BW	Beam	5	Waterborne	11	613	65
BS	Beam	5	Solvent-based	11	660	54
CW1/CW2	Column	4	Waterborne	8	691/677	64/61
CS1/CS2	Column	4	Solvent-based	8	655/680	51/55

provides a summary of the measured DFTs for the test specimens (Xu et al. 2020, 4). It can be seen that the measured average DFT values of the intumescent coatings were within ±10% of the designed value.

4.3 OBSERVATION OF THE LOCALIZED FIRE

4.3.1 Flame

The flames of the localized fire observed for the beam and column tests are shown in Figure 4.6. The flame height L_f can be estimated using the following equation (Heskestad 2016, 396–428):

$$L_f = -1.02D + 0.235\dot{Q}^{2/5} \tag{4.2}$$

where L_f and D are in m and $\dot{Q}$ is in kW. Substituting a value of D = 1.13 m and $\dot{Q}$ = 2.07 MW in Equation (4.2) gives a value of 3.82 m for the flame height, which agrees reasonably well with the observed flame heights of between 3.5 and 4.0 m.

It can be seen in Figure 4.6 that the flame slightly swayed to the north during the tests, which was due to the ventilation conditions in the laboratory. The north sway of the flame resulted in different thermal conditions for the north and south column specimens, with the north one effectively engulfed within the fire plume and the south one somewhat apart from the flames (Xu et al. 2020, 3–5).

4.3.2 Fire temperature distributions

A selection of measured gas temperature variations with time during localized fire tests around the beam and column specimens are shown in

(a)

(b)

Figure 4.6 Localized fire tests in progress. (a) Beam specimen test. (b) Column specimen test.

Figures 4.7 and 4.8, which were highly non-uniformly distributed and considerably fluctuated. Due to the ventilation conditions inside the test lab, the flame tilted to the north. As a result, the gas temperatures for south sections 5–8 of the beam specimens decreased at about 500 s, while those for the north sections (9–11) increased. Similarly, the flame surrounded the north column specimen on all surfaces, but not the south specimen, so the gas temperatures on the side and back surfaces of the north specimen (CW1, CS1) were more than 100°C higher than those of the south specimen (CW2, CS2) (Xu et al. 2020, 4–7).

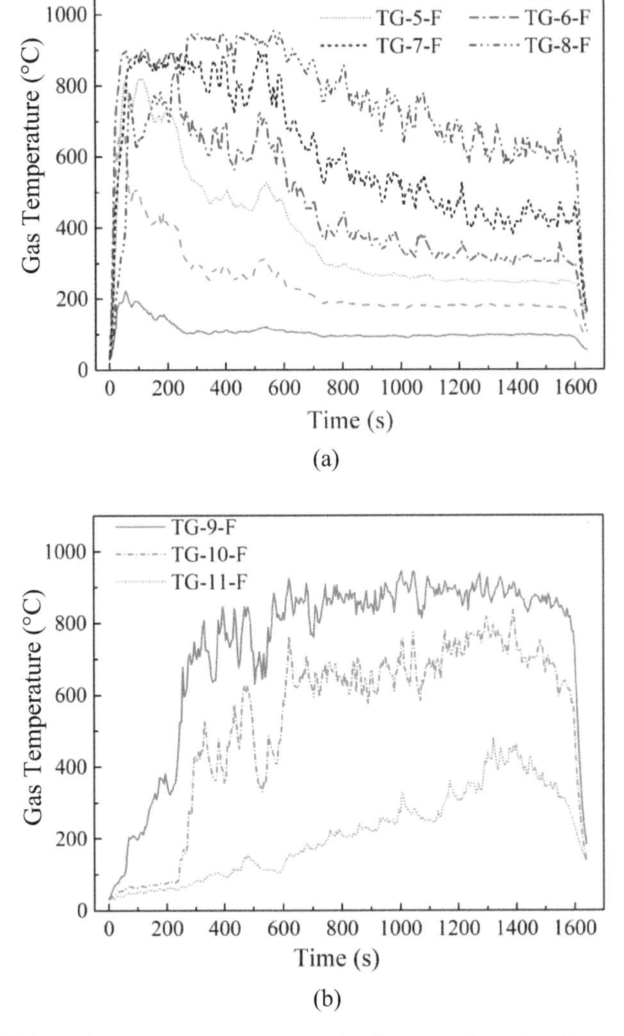

Figure 4.7 Selected gas temperatures at the front surface for beam specimens with waterborne intumescent coatings (BW)). (a) Gas temperatures for sections 3–8. (b) Gas temperatures for sections 9–11.

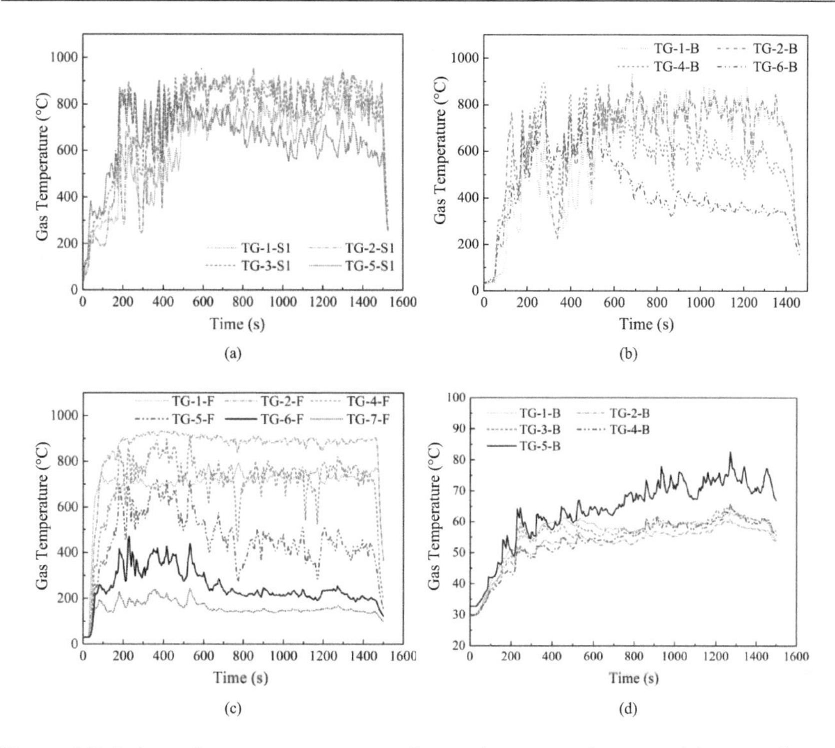

Figure 4.8 Selected gas temperatures for column specimens with waterborne intumescent coatings. (a) Side surface of north column CW1. (b) Back surface of north column CS1. (c) Front surface of south column CW2. (d) Back surface of south column CW2.

4.4 PERFORMANCE OF INTUMESCENT COATINGS

4.4.1 Reactions of intumescent coatings

Observations on the appearance of the intumescent coatings applied to the steel members after the localized fire tests show four regions: (1) fully expanded, (2) partially expanded, (3) melted but not expanded, and (4) virgin material, as illustrated in Figures 4.9 and 4.10 (Xu et al. 2020, 8). The intumescent coating is considered "fully expanded" if its expanded surface appearance is white, which indicates oxidation of the char, and bubbles are finely sized and reasonably uniformly distributed.

Region 1 was in general directly engulfed in the fire plume and the surrounding gas temperature was greater than 600°C to allow the intumescent coatings to undergo the full chemical reactions for expansion. Region 2 experienced expansion but did not fully expand, because the temperatures were insufficient for full chemical reactions, as in Region 1. Within this region, the recorded gas temperatures were between 300 and 600°C. In this region, the char surfaces were black, suggesting the presence of carbon on the char surface and an absence of significant char oxidation, with bubbles

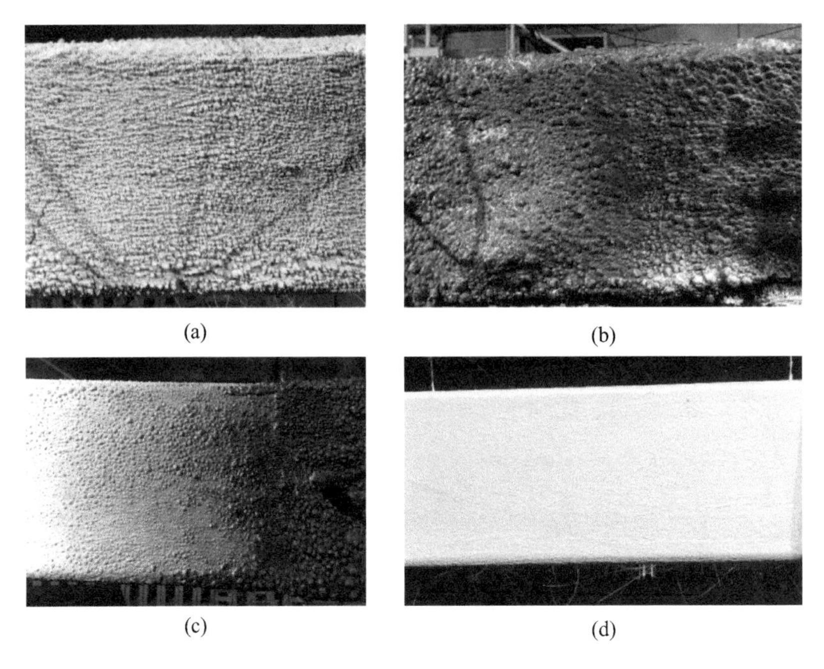

Figure 4.9 Typical surface appearances of four notional regions of waterborne intumescent coatings. (a) Full expansion region (Region 1). (b) Expanding region (Region 2). (c) Melting region (Region 3). (d) Virgin material (Region 4).

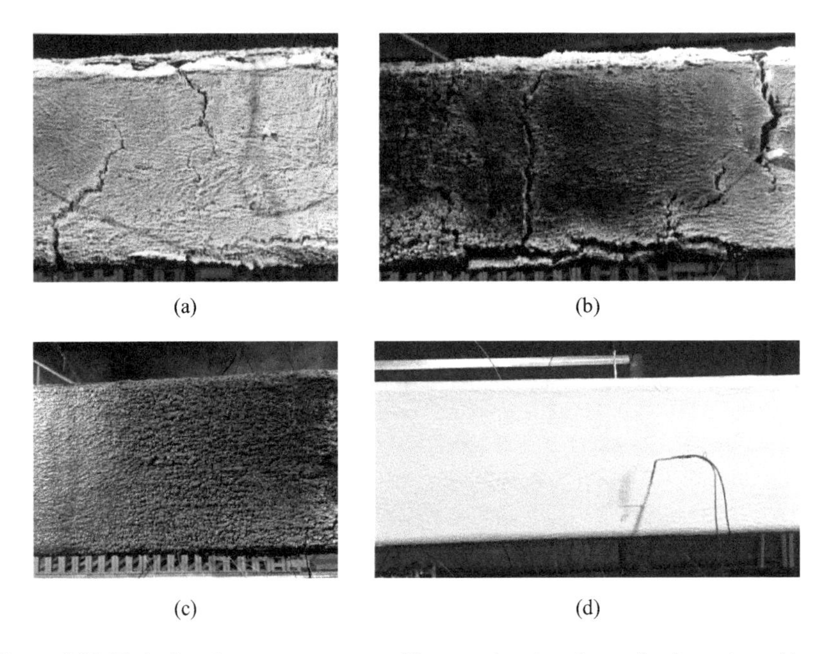

Figure 4.10 Typical surface appearances of four notional regions of solvent-based intumescent coatings. (a) Full expansion region (Region 1). (b) Expanding region (Region 2). (c) Melting region (Region 3). (d) Virgin material (Region 4).

of different sizes within the coating. Region 3 (melting region) displayed a few very small bubbles scattered on the coating surface, but no obvious coating expansion. The nearby fire temperature was higher than the melting temperature of intumescent coatings by about 150°C but lower than the temperatures required for the onset of expansion around 300°C (Griffin 2010, 249–277). In Region 4 (essentially virgin material), the surrounding gas temperatures were low (<100°C) and there was no indication of any melting or gasification processes.

The transition between different regions of the intumescent coatings after the localized fire tests was smooth, without any obvious particularities between the regions, and so these can be delineated only approximately. Different sizes of bubbles were observed at different expansion regions, where the coatings were at different stages of expansion.

As shown in Figure 4.9, the effectively expanded intumescent coatings in Region 1 had a continuous, smooth char, with small and evenly distributed white bubbles on the surface. This is similar to observations of fully expanded intumescent coatings in ISO 834 standard furnace tests (Xu et al. 2018, 31; Han et al. 2019, 5).

Figures 4.11–4.13 (Xu et al. 2020, 9–11) show the surface appearances of various regions for waterborne and solvent-based intumescent coatings

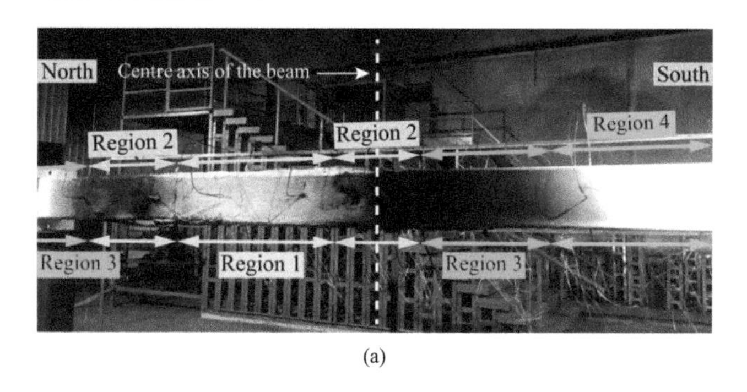

(a)

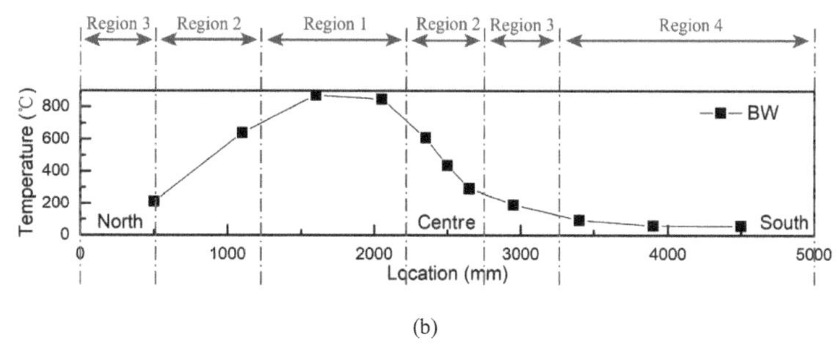

(b)

Figure 4.11 Distribution of four notional regions for waterborne coating on beam specimen and relation to average gas temperature distribution. (a) Distribution of the four regions on side surface I of beam specimen. (b) Approximate gas temperature distribution on side surface I of beam specimen during steady burning stage.

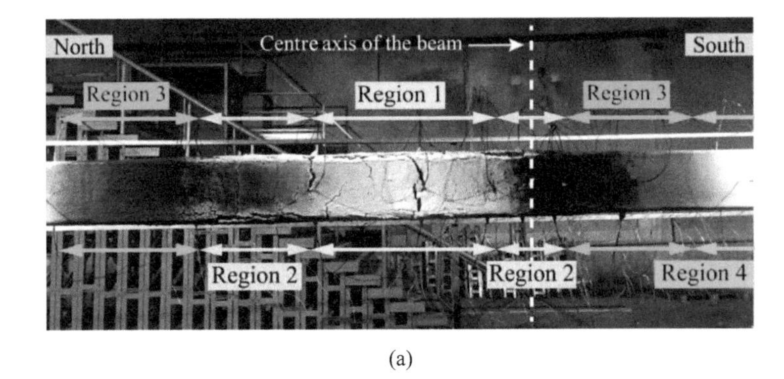

(a)

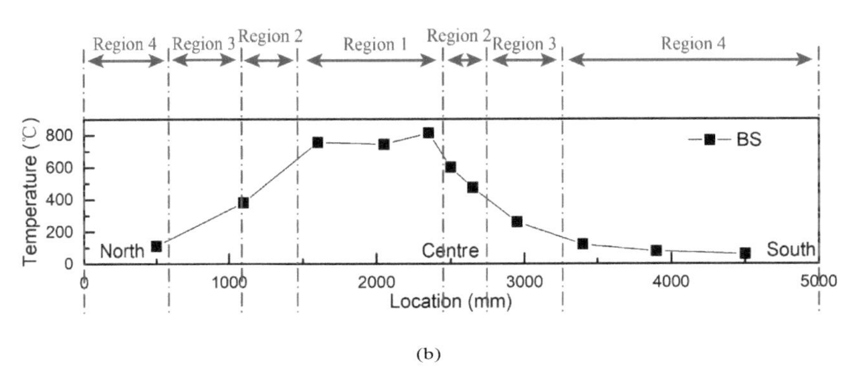

(b)

Figure 4.12 Distribution of four notional regions for solvent-based coating on beam specimen and relation to average gas temperature distribution. (a) Distribution of the four regions on side surface I of beam specimen. (b) Approximate gas temperature distribution on side surface I of beam specimen during steady burning stage.

corresponding to the distributions of gas temperatures along the length of the beam and column specimens respectively. As shown in Figure 4.13, the intumescent coatings on the two north column specimens (CS1 and CW1) reacted more fully (white char) than the two south ones (black char) due to higher gas temperatures as a result of the north shift of the flames.

In all the tests, the expansion ratios of the fully expanded regions (Region 1) were about 30 and 25 for the waterborne and solvent-based intumescent coatings respectively, which are the same as those in the uniform ISO standard furnace tests (Xu et al. 2018, 32–33). The expansion ratios of the expanding regions (Region 2) were approximately 20 and 15, respectively.

4.4.2 Cracking of intumescent coatings

Numerous cracks appeared on the surface of the coatings in all the test specimens, which accords with prior observations of intumescent coatings

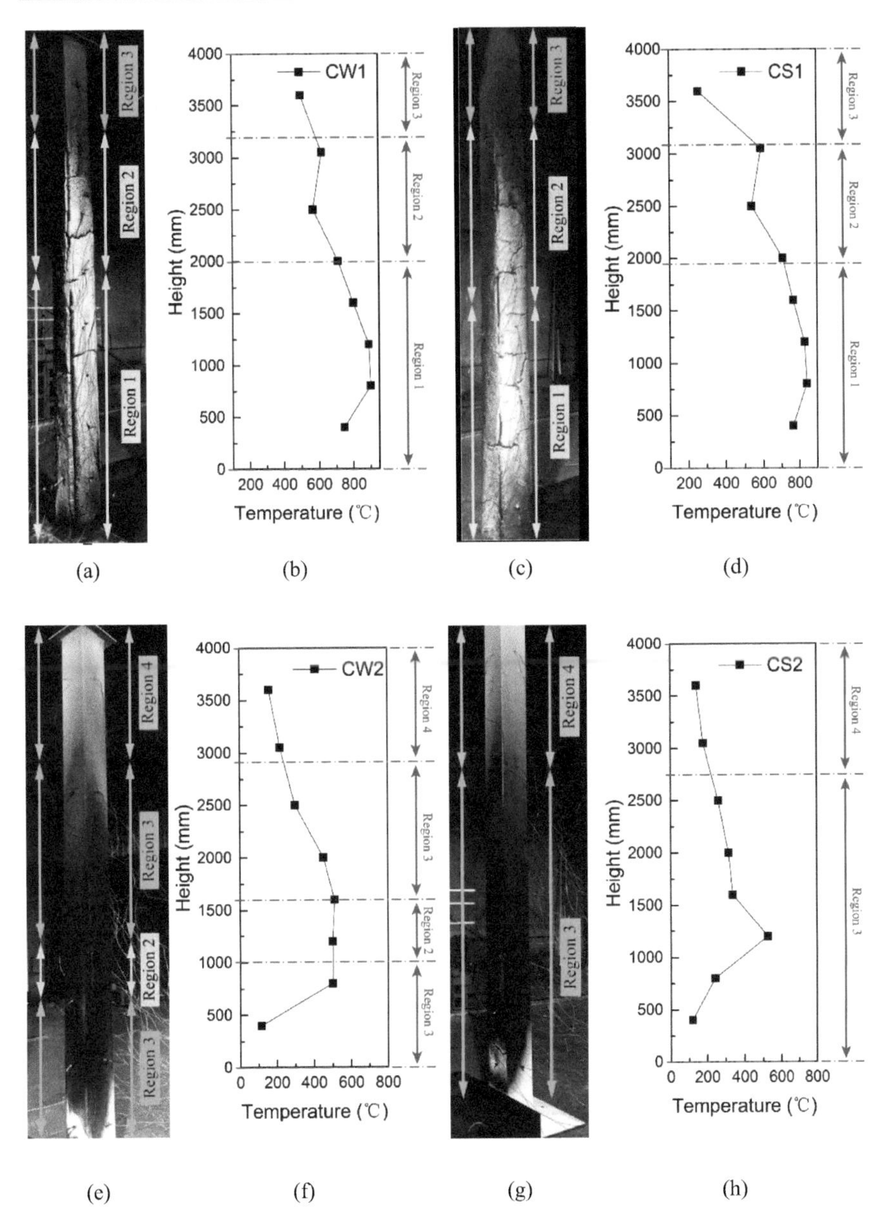

Figure 4.13 Distribution of different notional regions of intumescent coatings on column specimen and relation to average gas temperature distribution. (a) CW1, side surface 1. (b) Gas temperature distribution of CW1. (c) CS1, side surface 1. (d) Gas temperature distribution of CS1. (e) CW2, side surface 1. (f) Gas temperature distribution of CW2. (g) CS2, side surface 1. (h) Gas temperature distribution of CS2.

exposed to the uniform heating of the ISO standard fire and large space fire (Han et al. 2019, 3; Xu et al. 2018, 33). There were two types of cracks: scattered narrow cracks perpendicular to the edges, and long cracks along the specimens at the corners of the cross-sections, as shown schematically in Figure 4.14. The crack widths were typically between 2 and 6 mm, with only a few cracks at the corners exceeding 10 mm; these may be attributed to the intumescent coatings at the corners experiencing greater strains due to sharp changes in geometry at these locations.

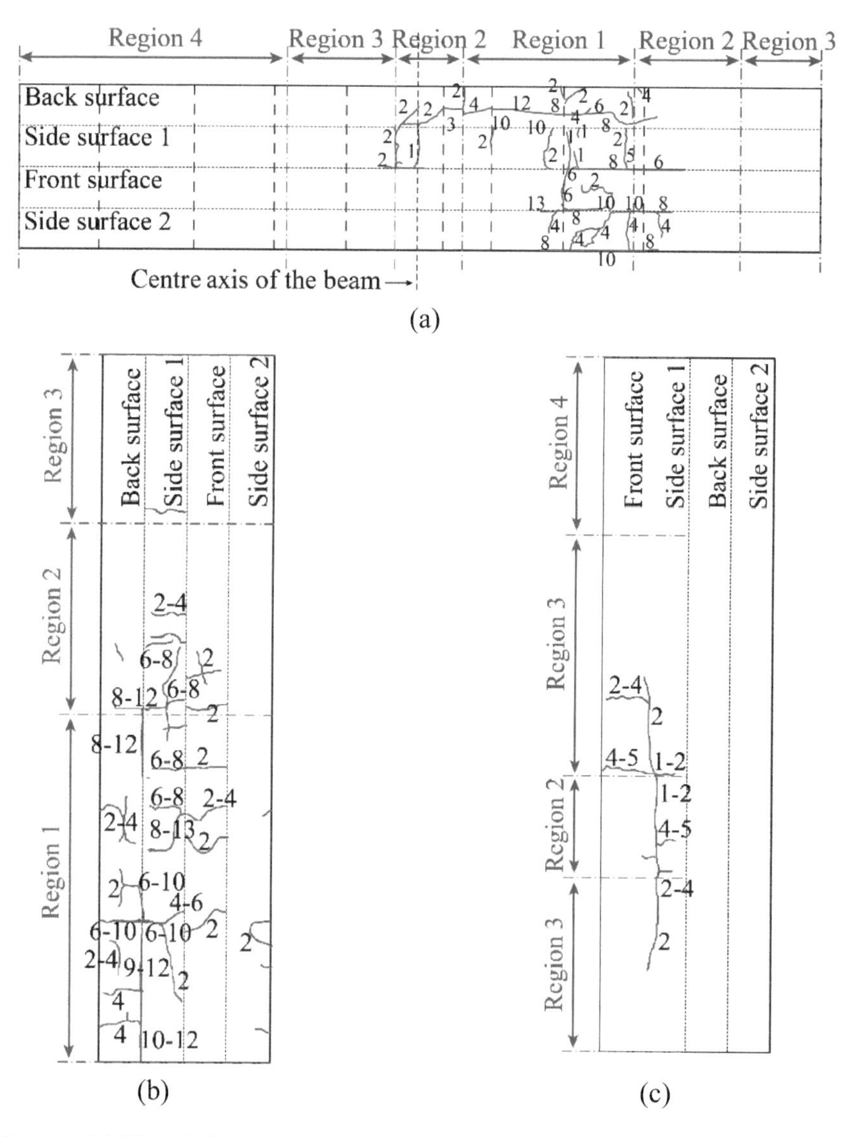

Figure 4.14 Typical distribution of cracks (numbers indicate crack width in mm). (a) BW. (b) CS1. (c) CW2.

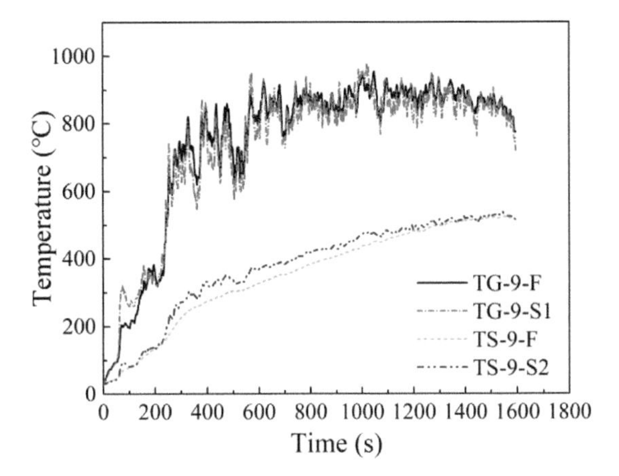

Figure 4.15 Comparison of measured steel temperatures at similar locations with and without cracking of coating for specimen BW.

Since most of the cracks were narrow and did not extended to the steel surface, their influence on heat transfer and steel temperature is considered to be slight. To confirm this assertion, Figure 4.15 compares the measured steel temperatures at a crack location (thermocouple TS-9-F, crack width 6 mm) and at the nearby location without a local crack (i.e. thermocouple TS-9-S1), both with similar adjacent gas temperatures and location facing to the fire source, for the beam specimen with waterborne intumescent coatings. It can be seen that the temperatures of the steel substrate protected with the intumescent coatings with and without cracking are similar.

4.4.3 Expansion of intumescent coatings

Figures 4.16–4.19 (Xu et al. 2020, 13–15) show the contours of the intumescent coating expansion ratio for both beam and column specimens correlating to the gas and steel temperature distributions during the steady burning periods of the tests (i.e. 600–1400 s). It can be seen that the full expansion regions (Region 1) coincide with the flame contacted area.

The average expansion ratios of the full expansion regions for the waterborne and solvent-based coatings were 29 and 21 respectively, and these were similar for both beam and column specimens. These expansion ratios agree with those for the same intumescent coatings under the ISO 834 standard furnace testing conditions (Xu et al. 2018, 32–33). This indicates that the same intumescent coatings exhibit similar behaviour at the fully expanded regions under spatially uniform and non-uniform heating conditions. This suggests that it may be possible to assume the same thermal properties of intumescent coatings obtained from standard furnace test conditions, which are likely to be easily available, to calculate protected steel temperatures under localized fire scenarios.

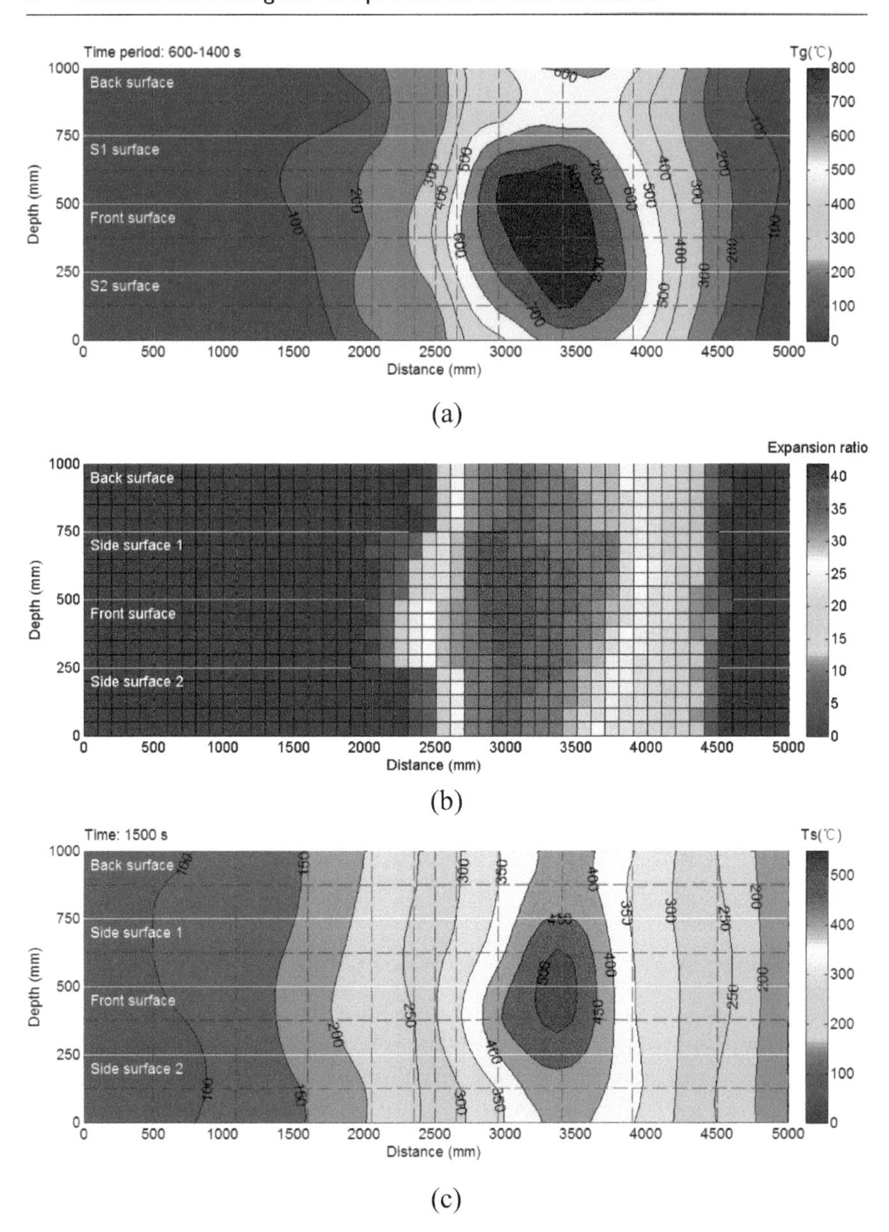

Figure 4.16 Measurements from beam specimen with waterborne intumescent coatings (specimen BW). (a) Gas temperature distribution. (b) Distribution of expansion ratios after fire. (c) Steel temperature distribution at 1500 s.

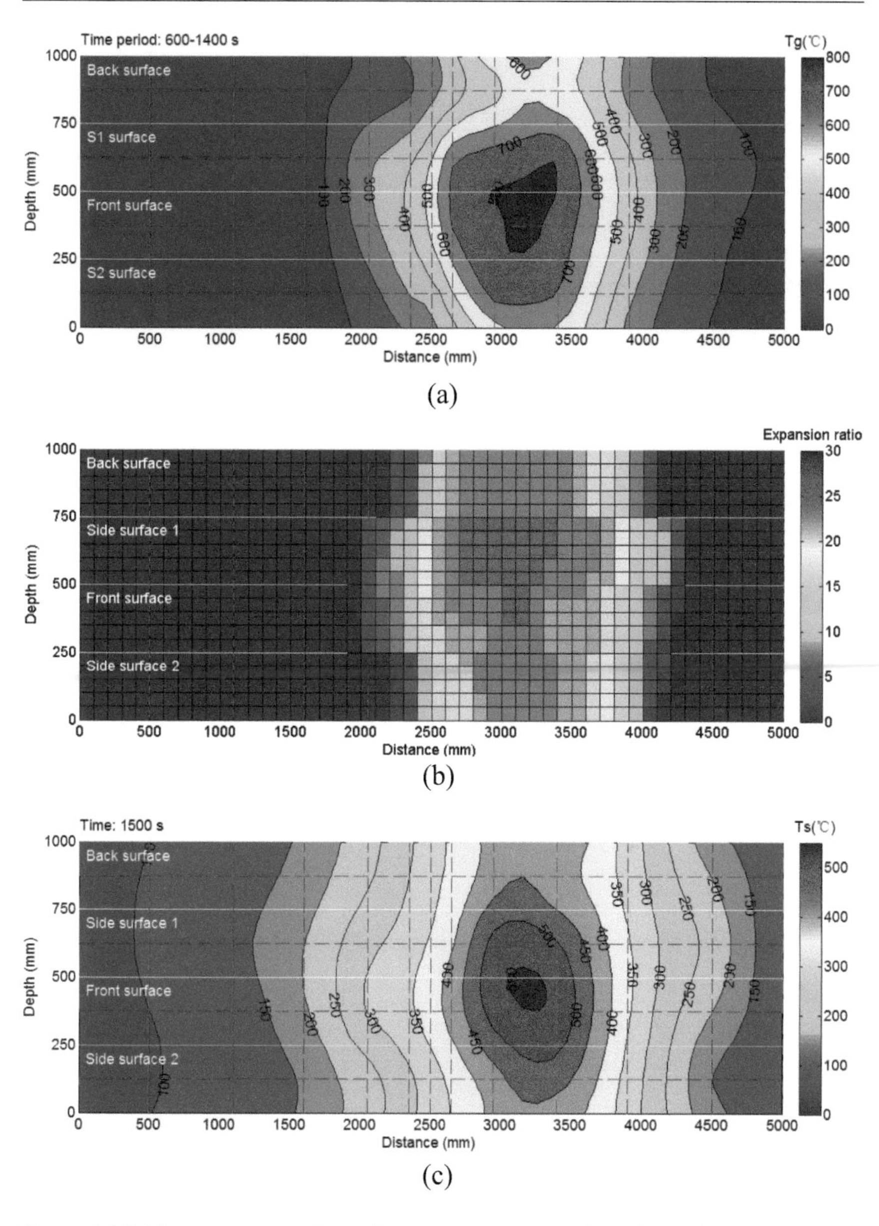

Figure 4.17 Measurements from beam specimen with solvent-based intumescent coatings (specimen BS). (a) Gas temperature distribution. (b) Distribution of expansion ratios after fire. (c) Steel temperature distribution at 1500 s.

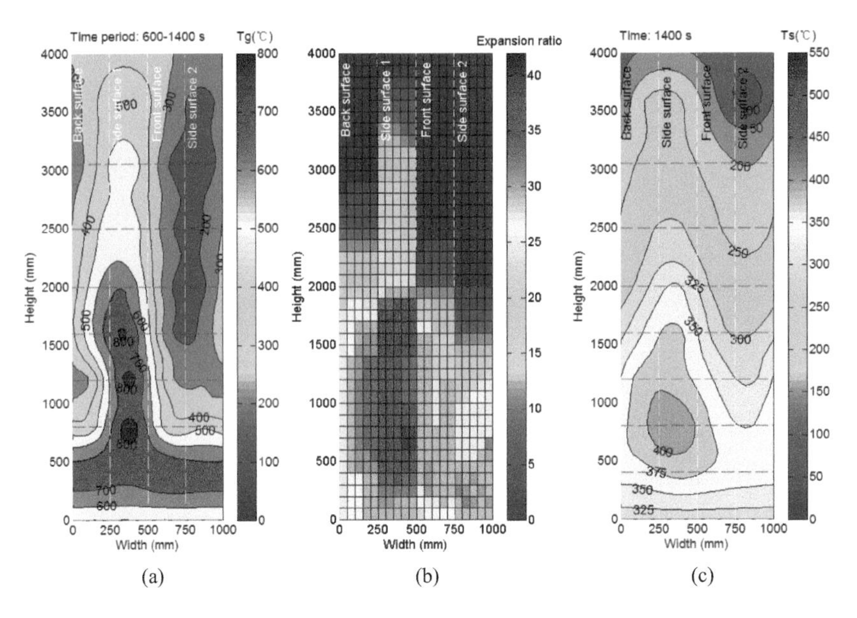

Figure 4.18 Measurements from column specimen with waterborne intumescent coatings (specimen CW1). (a) Gas temperature distribution. (b) Distribution of expansion ratio after fire. (c) Steel temperature distribution at 1500 s.

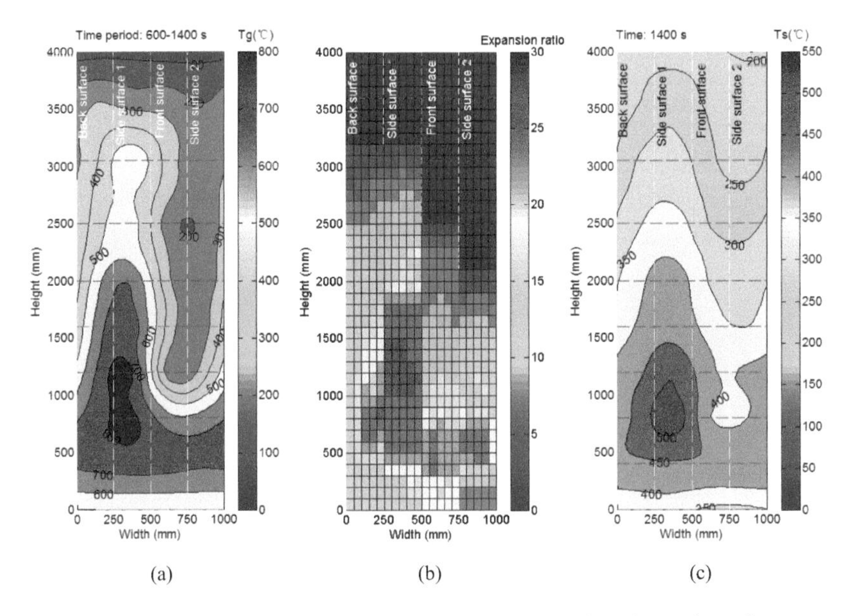

Figure 4.19 Measurements from column specimen with solvent-based intumescent coatings (specimen CS1). (a) Gas temperature distribution. (b) Distribution of expansion ratio after fire. (c) Steel temperature distribution at 1500 s.

4.4.4 Temperature-dependent effective thermal conductivity of intumescent coatings

Figure 4.20 (Xu et al. 2020, 17) shows the variations of temperature-dependent effective thermal conductivity with steel temperature, calculated by using Equation (2.5), for the fully expanded region for waterborne coating (BW-9-F) and solvent-based coating (BS-9-F), in comparison with the results obtained from the previous ISO furnace tests (Xu et al. 2018, 25–38) for the same intumescent coatings with the same DFT. The comparison shows that

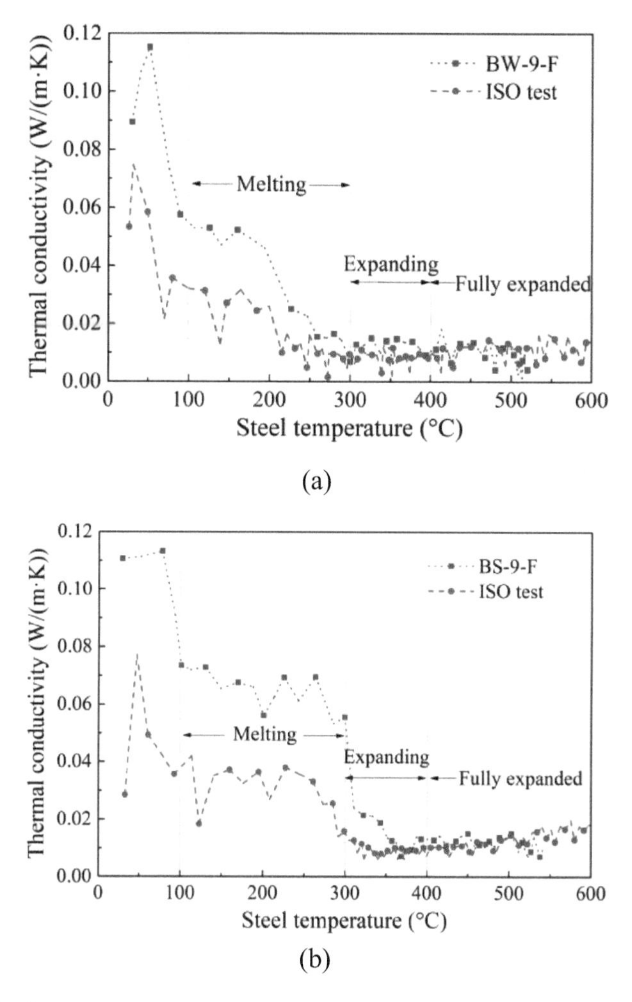

Figure 4.20 Comparison of temperature-dependent effective thermal conductivity with steel temperature curves for the full expanded region under localized fires and an ISO fire. (a) Waterborne intumescent coatings. (b) Solvent-based intumescent coatings.

the thermal conductivity of the intumescent coatings at the expanding and full expansion stages in the fully expanded region under localized fires were close to that determined with the furnace test under the ISO standard fire. However, the temperature-dependent effective thermal conductivity of the coatings at the melting stage under localized fires were different from that under the ISO standard fire. Nevertheless, it is not so important to achieve high accuracy in obtaining the protected steel substrate temperature below 300°C with accurate thermal conductivity of coatings for the safety of steel structures exposed to fire.

4.4.5 Validation of three-stage model for the thermal resistance of intumescent coatings under localized fires

To validate the assertion that protected steel temperatures under localized fire can be estimated by using the same three-stage model for the thermal resistance properties of intumescent coatings under uniform large space fire conditions, the estimated and measured steel temperatures in the different regions of the steel member specimens under localized fires were compared.

The measured steel temperatures refer to those at the front surfaces of sections 11, 6 and 9 for the beam specimen with a waterborne intumescent coating (BW) and sections 4, 10 and 8 for the beam specimen with a solvent-based intumescent coating (BS), which are located at the centre of the melting, expanding and full expansion regions, respectively.

In the estimation of steel temperatures, the thermal conductivities of intumescent coatings were assumed to be constant for the three stages of melting, expanding and full expansion, corresponding to the steel temperature ranges of 100–300°C, 300–400°C and above 400°C respectively (Han et al. 2019, 7). The constant effective thermal conductivities of the intumescent coatings for the fully expanded stage were adopted as the same as that obtained under the standard ISO 834 furnace test condition (Xu et al. 2018, 25–38), being 0.012 and 0.014 W/(m.K) respectively for the same type of waterborne and solvent-based intumescent coating with the same DFT of 0.6 mm. The ratio of the constant effective thermal conductivity at the melting stage of the coatings to that at the full expansion stage was taken as 4.0 (Xu et al. 2021, 2), and the ratio for the expanding stage was taken as 0.5 (Han et al. 2019, 8), as presented in Chapter 5.

Figure 4.21 (Xu et al. 2020, 18) shows that the estimated steel temperature results and the measured ones are in reasonable agreement at various regions of the beam specimen under localized fire. The relatively small differences between the estimated and measured steel temperatures, being less than 50°C for most locations, may be attributed to ignoring heat conduction within the steel members in the estimation.

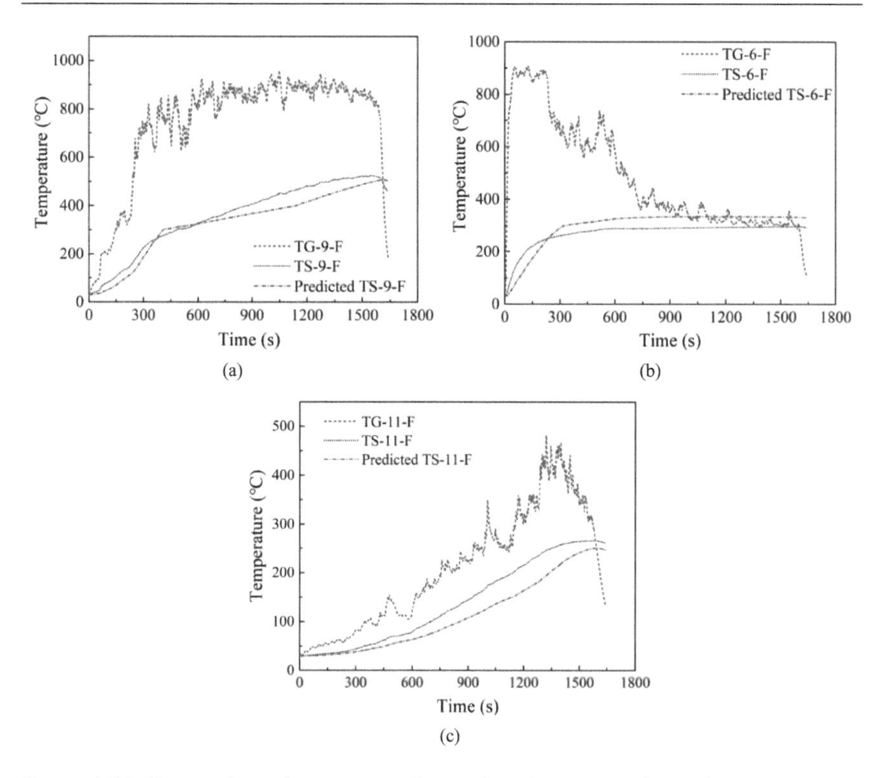

Figure 4.21 Comparison between estimated and measured steel temperatures for specimen BW. (a) Fully expanded region. (b) Expanding region. (c) Melting region.

4.5 SUMMARY

The main findings of this chapter can be summarized as follows.

1. Intumescent coatings at various locations of steel members behave differently due to the highly non-uniform heating conditions under localized fire. The behaviour of intumescent coatings can be classified into four notional categories: full expansion, partially expanded, melted but not expanded, and essentially unreacted. The steel temperature ranges in the regions of the four coating performance categories are approximately more than 400°C, 300–400°C, 100–300°C, and less than 100°C, respectively, which is in agreement with observations made under furnace tests with large space fire exposures.
2. The expansion ratios and the constant effective thermal conductivities of intumescent coatings in the full expansion regions under localized fire agree reasonably well with the results of fully expanded coatings from the furnace tests with uniform heating conditions.

3. Cracks, with widths in the range of 2–6 mm, may appear on the surfaces of intumescent coatings in the flame contact zone, which does not penetrate through the entire coating thickness to reach to the steel substrate surface. The effects of coating cracks on steel substrate temperature are not significant.
4. The three-stage thermal insulative properties of the same intumescent coatings obtained with uniform fire furnace tests can be reasonably applied to those under localized heating scenarios.

REFERENCES

Byström, Alexandra, Sjöström, Johan, Wickström, Ulf, Lange, David, and Veljkovic, Milan. 2014. "Large Scale Test on a Steel Column Exposed to Localized Fire." *Journal of Structural Fire Engineering* 5(2): 147–160. https://doi.org/10.1260/2040-2317.5.2.147.

de Silva, Donatella, Bilotta, Antonio, and Nigro, Emidio. 2019. "Experimental Investigation on Steel Elements Protected with Intumescent Coating." *Construction and Building Materials* 205: 232–244. https://doi.org/10.1016/j.conbuildmat.2019.01.223.

Griffin, G. J. 2010. "The Modeling of Heat Transfer across Intumescent Polymer Coatings." *Journal of Fire Sciences* 28(3): 249–277.

Han, Jun, Li, Guo-Qiang, Wang, Yong C., and Xu, Qing. 2019. "An Experimental Study to Assess the Feasibility of a Three Stage Thermal Conductivity Model for Intumescent Coatings in Large Space Fires." *Fire Safety Journal* 109: 102860. https://doi.org/10.1016/j.firesaf.2019.102860.

Heskestad, G. 2016. "Fire Plumes, Flame Height, and Air Entrainment." In *SFPE Handbook of Fire Protection Engineering*, 5th Edition, 396–428. New York.

Hostikka, S., Kokkala, M., and Vaari, J. 2001. *Experimental Study of the Localized Room Fires*.Espoo, Finland: VTT Technical Research Centre of Finland.

Xu, Qing. 2021. *Study on the Performance of Steel Structures Protected by Intumescent Coatings under Localized Fires*. Shanghai: Tongji University.

Xu, Qing, Li, Guo-Qiang, Jiang, Jian, and Wang, Yong C. 2018. "Experimental Study of the Influence of Topcoat on Insulation Performance of Intumescent Coatings for Steel Structures." *Fire Safety Journal* 101: 25–38. https://doi.org/10.1016/j.firesaf.2018.08.006.

Xu, Qing, Li, Guo-Qiang, and Wang, Yong C. 2021. "A Simplified Method for Calculating Non-uniform Temperature Distributions in Thin-Walled Steel Members Protected by Intumescent Coatings under Localized Fires." *Thin-Walled Structures* 162: 107580. https://doi.org/10.1016/j.tws.2021.107580.

Xu, Qing, Li, Guo-Qiang, Wang, Yong C., and Bisby, Luke. 2020. "An Experimental Study of the Behavior of Intumescent Coatings under Localized Fires." *Fire Safety Journal* 115: 103003. https://doi.org/10.1016/j.firesaf.2020.103003.

Zhang, Chao, and Li, Guo-Qiang. 2012. "Fire Dynamic Simulation on Thermal Actions in Localized Fires in Large Enclosure." *Advanced Steel Construction* 8(2): 124–136.

Chapter 5

Hydrothermal ageing effects on the insulative properties of intumescent coatings

When specifying intumescent coatings for the fire protection design of steel structures, the following assumptions are employed:

1. The type and thickness of the intumescent coating are correctly specified;
2. The intumescent coating is correctly applied;
3. The fire protection performance of the intumescent coating does not degrade over time.

Assumptions (1) and (2) may be fulfilled in practice through strict management, while assumption (3) is related to the durability of intumescent coatings. Since intumescent coatings are chemically reactive products, their fire protection properties may degrade after prolonged exposure to environmental conditions. The capacity for fire resistance of the protected steel structures decreases with the degradation of the insulative properties of intumescent coatings, which limits the period of the coatings' design life. So, the reliability of intumescent-coating-protected steel structures in their design life has been of wide concern to code authorities, fire departments, engineers and coating manufacturers.

This chapter explains why the insulative properties of intumescent coatings degrade after exposure to hydrothermal environmental conditions and presents quantitative information on the insulative degradation of intumescent coatings with service time.

5.1 AGEING MECHANISM OF INTUMESCENT COATINGS IN A HYDROTHERMAL ENVIRONMENT

An intumescent coating is typically composed of three active components (an acid source, blowing agent and charring agent) bound in a binder polymer. When exposed to heat, several reactions (melting, intumescence, char

DOI: 10.1201/9781003287919-5

formation and char oxidation) occur in series and in parallel (Mamleev et al. 1998, 1524; Jimenez et al. 2006, 17). These reactions cause the coating to swell and to form a porous flame retardant char with low thermal conductivity. It is essential that different components in intumescent coatings show a suitable matching thermal behaviour and the order and timing of chemical and physical processes happen in an appropriate sequence (Zhang et al. 2012, 52; Wang et al. 2019, 783). Therefore, ingredients and their mixture are meticulously designed to obtain an intumescent coating char with superior performance.

When exposed to hydrothermal environmental conditions, the hydrophilic components of an intumescent coating move to the surface of the coating and can be dissolved by moisture in the air, which can destroy the intended chemical reactions of these components with others and deter formation of the desired effective intumescent char. The consequence of this reduces the expansion of the coating and increases the effective thermal conductivity. To investigate the migration of chemical components in intumescent coatings, Wang et al. (2013, 176–179) conducted a Fourier transform infrared spectroscopy (FTIR) test and an X-ray photoelectron spectroscopy (XPS) test on samples extracted from the surface layer of aged intumescent coatings.

5.1.1 Test preparation

Each specimen prepared for FTIR and XPS tests was made of a 10 mm thick steel plate coated with 1 mm dry film thickness intumescent coatings on the top side. Two different types (type-A and type-U) of intumescent coating were applied with two different manufacturers operating in the Chinese market. The principal components of type-U and type-A intumescent coating are APP-MEL-DPER (aided with zinc borate) and APP-MEL-PER, respectively; the acid resins of type-U and type-A intumescent coatings are ethylene benzene-acrylic and single component acrylic, respectively. Before the FTIR and XPS tests, ageing tests were performed on these specimens for four different cycles of accelerated ageing, including 0 cycles (no ageing), 11 cycles (simulating five years in service), 21 cycles (simulating ten years in service) and 42 cycles (simulating 20 years in service). Detailed information of the ageing tests will be given in Section 5.2.

5.1.2 FTIR test results

The FTIR test results are presented in Figure 5.1. For type-A coatings (Figure 5.1a), the troughs at 3321 cm^{-1}, 3176 cm^{-1}, 2957 cm^{-1} and 1668 cm^{-1} indicate N-H bonds contained in MEL and APP, O-H bonds in PER, C-H bonds in acrylic acid resin and PER, and C=O bonds in acrylic acid resin respectively. The trough at 1428 cm^{-1} is the overlapping peak of the

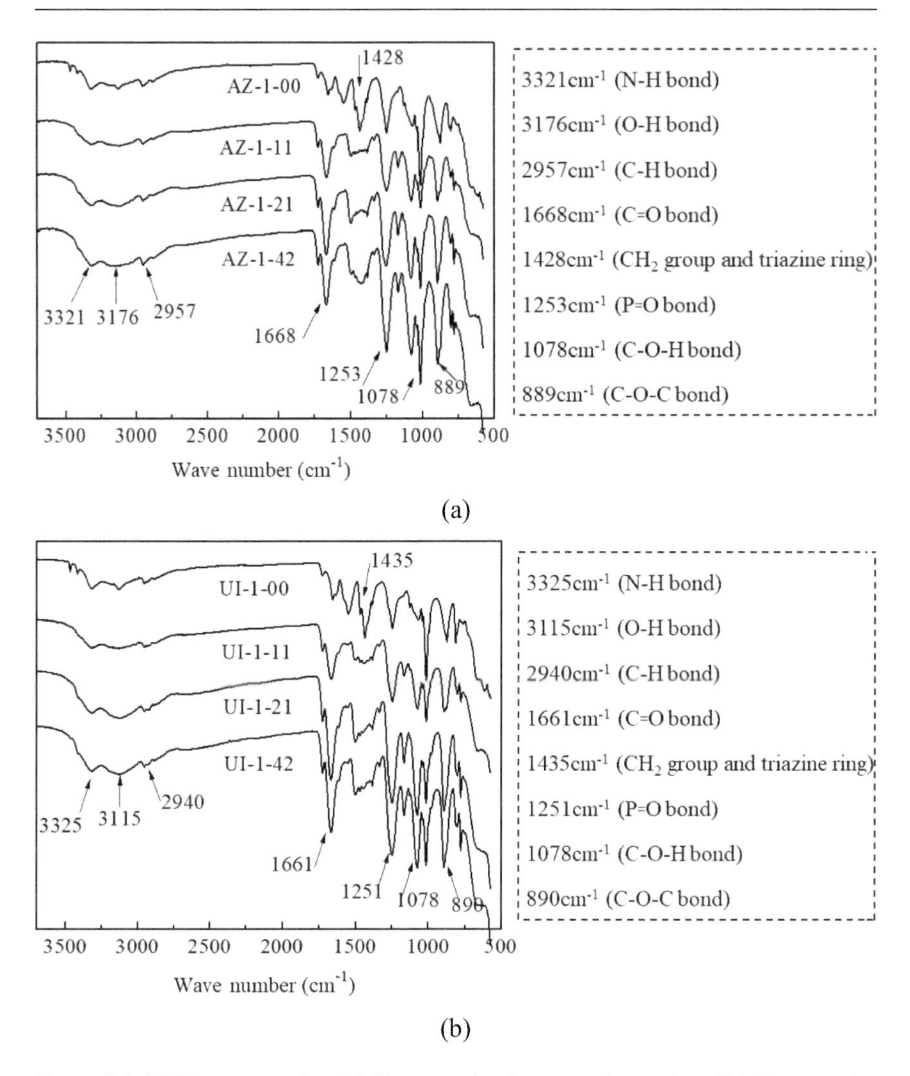

Figure 5.1 FTIR test results. (a) Test results for type-A coating. (b) Test results for type-U coating.

absorption peak of the CH_2 group contained in acrylic acid resin and PER, and the absorption peak of triazine rings which are the main structure of MEL. The troughs at 1253 cm^{-1}, 1078 cm^{-1}, 1013 cm^{-1} and 889 cm^{-1} indicate P=O bonds in APP, C-O-H bonds in PER, C-O-C bonds in acrylic acid resin and triazine rings in MEL respectively. Similar wave numbers of these bonds can be observed in Figure 5.1(b) for type-U coating.

It is clear from Figure 5.1 that the absorption peak of the above mentioned different chemical bonds contained in PER and APP are enhanced with the increasing number of cycles of ageing. This indicates that PER and APP migrated from within the coating to the surface of the coating under the development of ageing.

Compared to type-A coating without ageing, the absorption peak at 1428 cm^{-1} of type-A coating after 11 cycles of ageing was weakened, whereas the width of the peak increased. This indicates that the polymer binder degraded under the effect of water and oxygen and some of the CH_2 groups were oxidated into C=O groups. This enhanced the absorption peak of C=O bonds at 1668 cm^{-1} with the increasing number of cycles of ageing. Degradation of the polymer binder (acrylic acid resin) was present during the whole process of ageing. The absorption peaks at 1428 cm^{-1} and 889 cm^{-1} were also enhanced, indicating migration of MEL from within the coating to the surface of the coating after different cycles of ageing.

It can be seen from the above analysis that the degradation of the polymer binder (acrylic acid resin) and the migration of the flame retardant system (APP-MEL-PER) happened at the same time during the process of ageing, which resulted in the reduced fire protective properties of the intumescent coatings.

Although the FTIR test results can only be used to gain a qualitative understanding of the effects of ageing, they can still give some indication of the extent of ageing in service.

In practical application, when examining the effects of ageing on intumescent coatings, if the on-site FTIR test shows little change in the absorption peaks, then there is high confidence that the effects of ageing are minimal.

5.1.3 XPS test results

The XPS test gives information on the amount of chemical elements being examined. For example, Figure 5.2 presents the amounts of carbon and nitrogen existent on the surface layer of both types of intumescent coatings after different cycles of ageing.

The element of carbon (C) is contained in MEL($C_3H_6N_6$) which acts as the blowing agent and in DPER/PER ($C(CH_2OH)_4$) acting as the charring agent. The element of nitrogen (N) is contained in MEL and APP($(NH_4)_{n+2}P_nO_{3n+1}$) which acts as the catalytic agent. Table 5.1 lists the percentage of carbon and nitrogen elements obtained from the XPS tests for representative samples of both types of intumescent coatings.

It can be seen from Table 5.1 that compared to specimens without ageing, the content of elements C and N on the surface layer of type-A and type-U coating specimens increased with the increasing number of ageing cycles. The change in element N is much more sensitive to the change in element C.

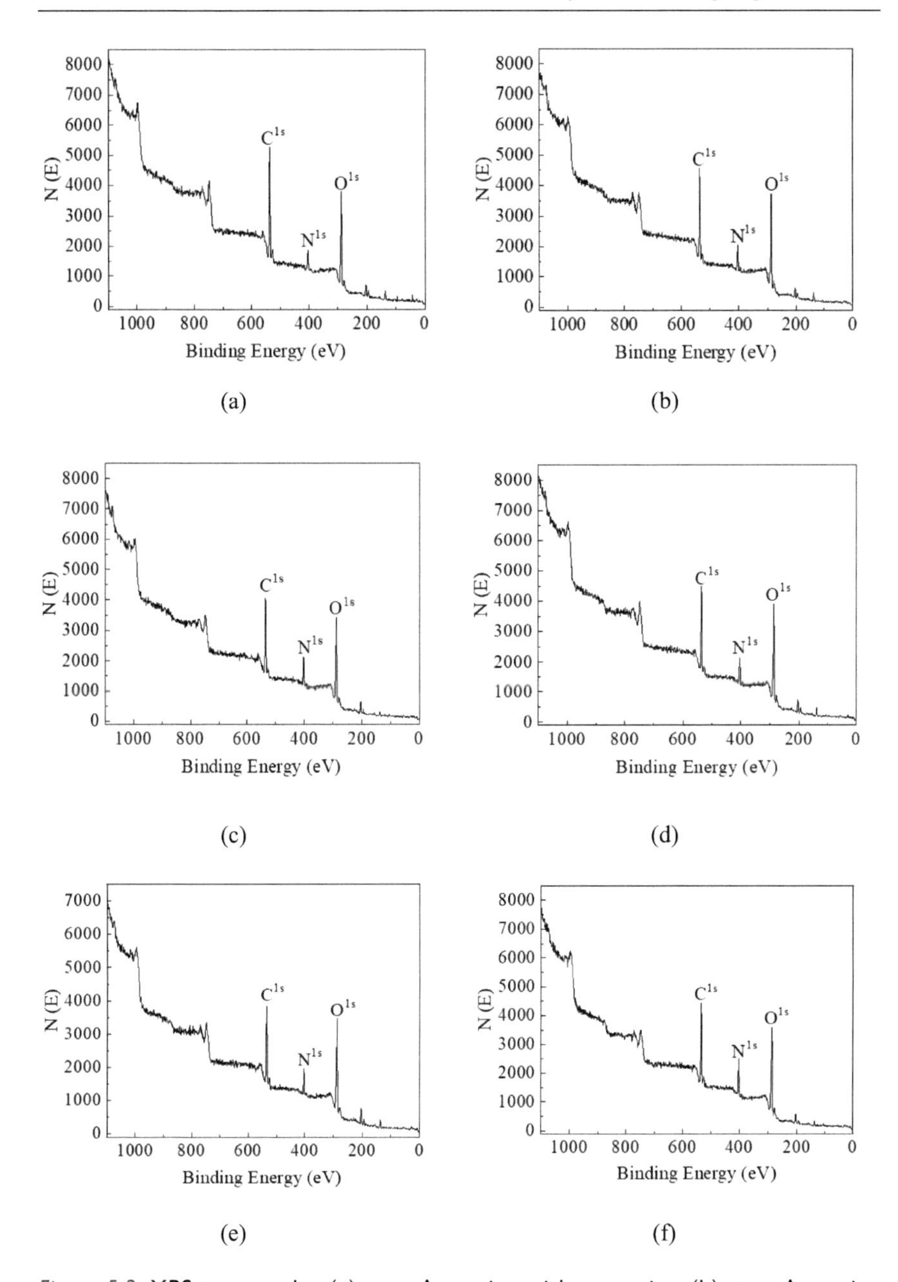

Figure 5.2 XPS test results. (a) type-A coating without ageing. (b) type-A coating with 21 cycles of ageing. (c) type-A coating with 42 cycles of ageing. (d) type-U coating without ageing. (e) type-U coating with 21 cycles of ageing. (f) type-U coating with 42 cycles of ageing.

Table 5.1 Content of elements carbon and nitrogen

	Element content					
	Type-A coating with ageing cycles			*Type-U coating with ageing cycles*		
Element	*0 cycles*	*21 cycles*	*42 cycles*	*0 cycles*	*21 cycles*	*42 cycles*
C	61.6	64.1	65.5	63.7	63.8	64.9
N	7.9	8.7	10.5	9.9	10.1	13.4

5.2 DEGRADATION OF INTUMESCENT COATINGS DUE TO AGEING IN A HYDROTHERMAL ENVIRONMENT

To quantitatively investigate the degradation of intumescent coatings in a hydrothermal environment, steel plate specimens protected with these coatings, subjected to different cycles of accelerated ageing, were tested in the standard fire to evaluate the variation of the thermal insulation properties with the evolution of ageing.

5.2.1 Test specimens with intumescent coatings

A total of 56 specimens were tested. Each specimen was made of a 16 mm thick steel plate coated with a 1 or 2 mm dry film thickness (DFT) of intumescent coatings on all sides. Of all the specimens, 16 were protected by type-U intumescent coating (to be referred to as type-U specimens) and 40 were applied with a type-A intumescent coating (to be referred to as type-A specimens). The principal components of type-U and type-A intumescent coatings were presented in Section 5.1.

For the 16 type-U specimens, four replicate tests were performed for the following four cycles of accelerated ageing: 0 cycles (no ageing), 11 cycles (simulating five years in service), 21 cycles (simulating ten years in service) and 42 cycles (simulating 20 years in service). All the specimens were coated with 1 mm DFT. For the 40 type-A specimens, 20 were coated with 1 mm DFT and the other 20 with 2 mm DFT. Also four replicate tests were performed for the following five cycles of accelerated ageing: 0 cycles (no ageing), four cycles (simulating two years in service), 11 cycles (simulating five years in service), 21 cycles (simulating ten years in service) and 42 cycles (simulating 20 years in service). For all the tested specimens, the substrate steel plate measured 200 mm by 270 mm by 16 mm thick. A primer was applied to the steel surface first to act as an aid to adhesion of the intumescent coating, then followed by different layers of the same coating to achieve the desired DFT. No topcoat was applied to any of the specimens. For each specimen, the DFT was measured and recorded before the accelerated ageing test. Three thermocouples were embedded in each steel plate. The main specimen parameters are listed in Table 5.2 and the specimen dimensions are shown in Figure 5.3, where d is the intumescent DFT.

Table 5.2 Main parameters of test specimens

Coating type	*Coating DFT (mm)*	*No. of cycles of accelerated ageing*	*Simulated time in service (years)*	*Specimen ID*
U	1	0	0	UI-1-00-*i* (*i*=1, 2, 3, 4)
		11	5	UI-1-11-*i* (*i*=1, 2, 3, 4)
		21	10	UI-1-21-*i* (*i*=1, 2, 3, 4)
		42	20	UI-1-42-*i* (*i*=1, 2, 3, 4)
A	1	0	0	AZ-1-00-*i* (*i*=1, 2, 3, 4)
		4	2	AZ-1-04-*i* (*i*=1, 2, 3, 4)
		11	5	AZ-1-11-*i* (*i*=1, 2, 3, 4)
		21	10	AZ-1-21-*i* (*i*=1, 2, 3, 4)
		42	20	AZ-1-42-*i* (*i*=1, 2, 3, 4)
A	2	0	0	AZ-2-00-*i* (*i*=1, 2, 3, 4)
		4	2	AZ-2-04-*i* (*i*=1, 2, 3, 4)
		11	5	AZ-2-11-*i* (*i*=1, 2, 3, 4)
		21	10	AZ-2-21-*i* (*i*=1, 2, 3, 4)
		42	20	AZ-2-42-*i* (*i*=1, 2, 3, 4)

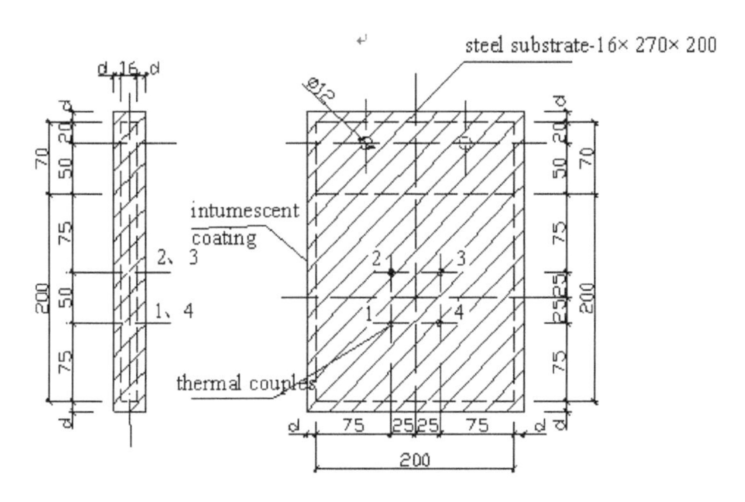

Figure 5.3 Specimen dimensions.

5.2.2 Hydrothermal ageing tests

Intumescent coating ageing is an extremely complicated process of physical and chemical interactions between the chemical components of intumescent coatings and the external environment. Whilst it would be ideal to carry out a real time ageing test, this process would be extremely long, running into many tens of years, and be hard to conduct systematic comparative tests. An alternative is to conduct an accelerated ageing test, in which a real environmental condition over a long period of time is represented by a short cyclic exposure to a concentrated dosage of some environmental condition.

The accelerated ageing test was performed according to the European guideline EAD 350402-00-1106 (EOTA 2017). In this guidance, four types of environmental exposure are simulated: (a) type X for all conditions; (b) type Y for internal and semi-exposed conditions; (c) type Z1 for internal conditions which have a temperature above 0°C and a high humidity; and (d) type Z2 for internal conditions that have a temperature above 0°C but different humidity conditions from class Z1. In this study (Wang et al. 2013, 169), type Z1 environmental exposure was adopted, simulating the more severe exposure condition of application around coastal areas.

For the type Z1 condition, each cycle of exposure is specified as:

8 h at (40±3)°C and (98±2)%RH;
16 h at (23±3)°C and (75±2)%RH.

According to EAD 350402-00-1106 (EOTA 2017), 21 cycles of accelerated ageing is equivalent to ten years in service under real conditions. Based on this correlation, 0, 4, 11 and 42 cycles of accelerated ageing correspond to fresh use, 2 years, 5 years and 20 years in service respectively.

After the accelerated ageing test but before the fire test, the specimens were checked for their coating surface appearance, as shown in Figures 5.4 and 5.5.

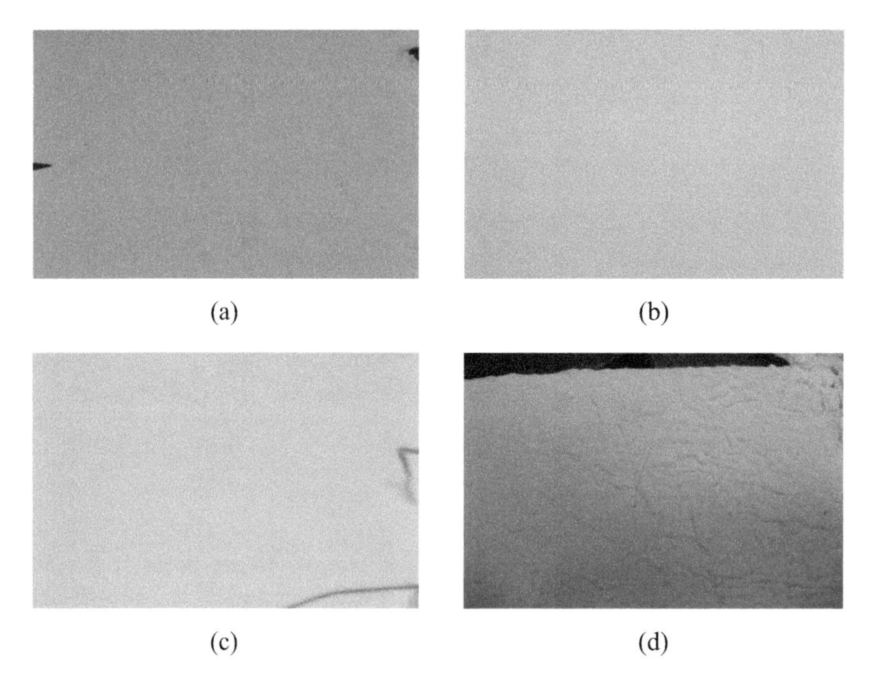

(a) (b) (c) (d)

Figure 5.4 Type-U coating appearance after different cycles of a hydrothermal ageing test. (a) UI-1-00. (b)UI-1-11. (c) UI-1-21. (d) UI-1-42.

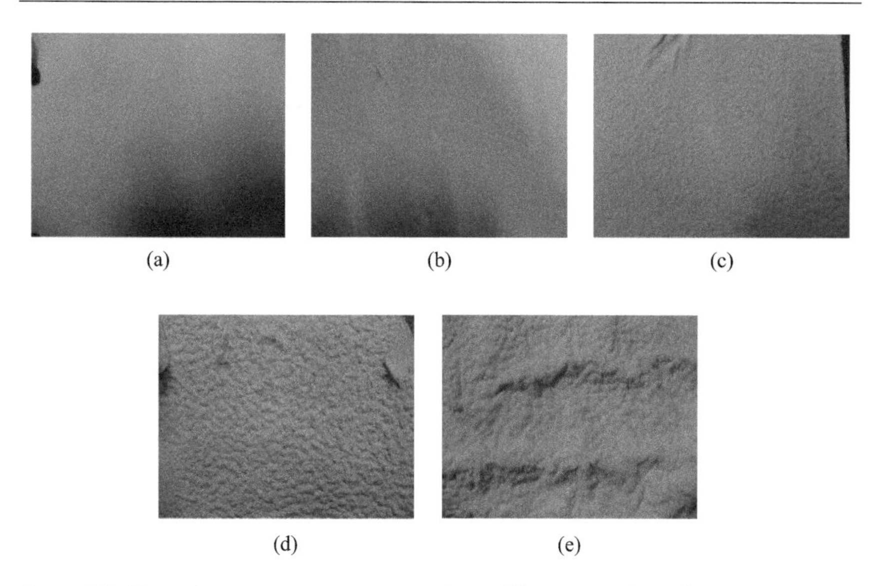

Figure 5.5 Type-A coating appearance after different cycles of a hydrothermal ageing test. (a) AZ-1-00. (b) AZ-1-04. (c) AZ-1-11. (d) AZ-1-21. (e) AZ-1-42.

Type-U specimens did not appear to suffer any change in appearance after 11 and 21 cycles of hydrothermal ageing tests (Figure 5.4b, c). After 42 cycles, wrinkles can be clearly seen (Figure 5.4d). In contrast, type-A specimens experienced noticeable changes in appearance after every accelerated ageing test. After only four cycles, the surface of type-A specimens appeared uneven (Figure 5.5b). After 11 cycles of an accelerated ageing test, bumps appeared on the surface (Figure 5.5c), and after 21 and 42 cycles, very large bumps appeared (Figure 5.5d, e). As will be seen in the following fire test, there is a strong link between the surface appearance and fire protection performance for the intumescent coating.

5.2.3 Fire tests

After the specimens were subjected to hydrothermal ageing tests as described in the previous section, they were placed in a furnace, as shown in Figure 5.6, and exposed to fire (Wang et al. 2013, 171). The furnace temperature was measured by four thermocouples and regulated according to the ISO 834 standard temperature–time relationship (ISO 1975).

Four specimens were tested together in the furnace, as shown in Figure 5.7. The steel temperature of the specimens was measured by three thermocouples embedded in the steel plate and recording was made every minute continuously. Four observation holes were placed on the furnace door to enable the fire tests to be observed. Each test was continued until the steel temperature reached 700°C.

Figure 5.6 Furnace of the fire test. (a) Specimens hung on steel beams. (b) Specimens laid flat on steel beams.

Figure 5.7 Specimens in the furnace.

When exposed to fire, intumescent coatings for all specimens underwent melting, expansion, char formation and char degradation due to oxidation. Depending on the composition of the chemical components and the fire exposure conditions, these reactions may happen in sequence or together. Type-A intumescent coatings began to expand earlier than type-U. In the intumescence (expansion) stage, bubbles that appeared on the surface of type-A intumescent coatings were much larger than those on the surface of type-U.

Intumescent coatings for both types are highly "engineered" to pass the standard fire resistance test when freshly applied. Hydrothermal ageing causes some chemical components in the intumescent coatings to migrate to the surface, altering the chemical reactions. In the intumescence stage, the blowing agent in the intumescent coating decomposes to produce gas; the gas is trapped within the molten matrix and causes the coating to expand.

From the pictures taken of the specimens through the observation holes on the furnace door, many bubbles appeared during the intumescence stage on both types of intumescent coating after zero or four cycles of hydrothermal ageing tests, as shown in Figures 5.8(a) and 5.9(a). This means a large amount of gas was produced due to decomposition of the blowing agent. However, after 11 and 21 cycles of hydrothermal ageing tests, the number of bubbles decreased drastically and the bubble distribution was much less uniform, as shown in Figures 5.8(b) and 5.9(b). After 42 cycles, bubbles were almost non-existent, as shown in Figures 5.8(c) and 5.9(c).

The observed phenomena for type-A specimens with both 1 and 2 mm DFTs were generally similar.

The most important parameters that directly reflect the fire protection performance of intumescent coatings are the final expanded thickness and internal structure of the char (Wang 2011, 75). The expanded thickness decreased with increasing cycles of the ageing test for both types of intumescent coatings, as shown as Figure 5.10. In addition, the integrity and consistency of the char for the specimens after the ageing test become poor, especially after 42 cycles of ageing.

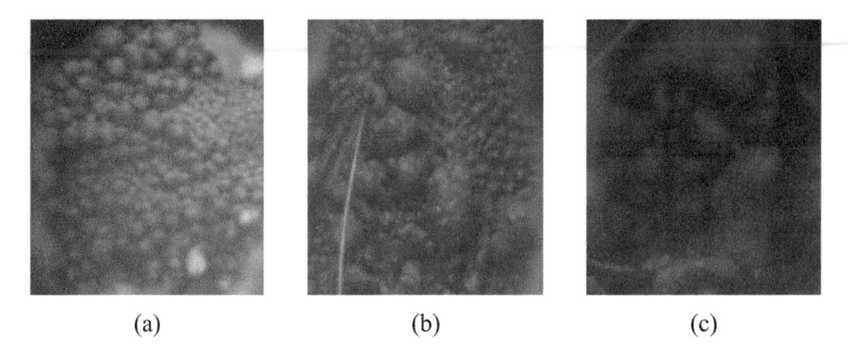

(a) (b) (c)

Figure 5.8 Bubble appearance on the surface of type-A specimens. (a) AZ-1-00. (b) AZ-1-21. (c) AZ-1-42.

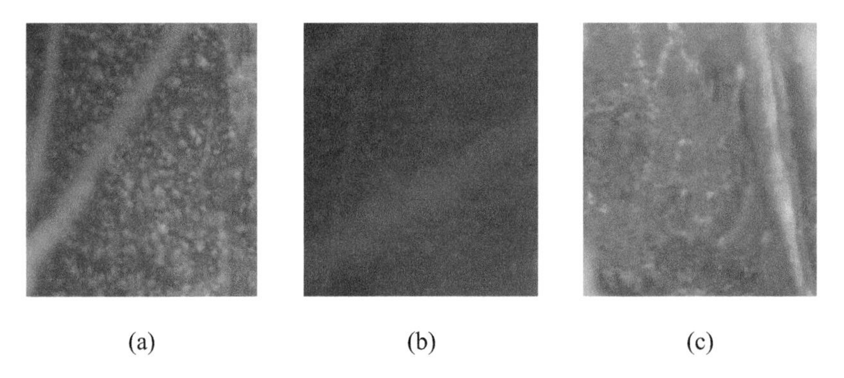

(a) (b) (c)

Figure 5.9 Bubble appearance on the surface of type-U specimens. (a) UI-1-00. (b) UI-1-21. (c) UI-1-42.

Figure 5.10 Cross-sectional view of expanded intumescent char after fire tests for specimens with different cycles of accelerated ageing. (a) AZ-2-00. (b) AZ-2-42. (c) AZ-1-04. (d) AZ-1-42. (e) UI-1-11. (f) UI-1-42.

Table 5.3 lists the measured DFTs, the measured final thickness of coating char and the expansion ratios of coating char to DFT for the different specimens. The expansion ratio is plotted in Figure 5.11 as a function of the number of ageing cycles. It can be seen that the expansion ratio decreases considerably after only a few cycles of an ageing test. After 21 cycles of an ageing test, simulating ten years in service, the expansion ratios of all three groups of intumescent coatings were about 60% of those without ageing, while after 42 cycles of an ageing test, the expansion ratio was about one-third of that without ageing.

Table 5.3 Expansion ratios of coating for various specimens

Specimen	*Initial thickness (mm)*	*Final thickness (mm)*	*Expansion ratio*
AZ-1-00	1.02	47.00	46.08
AZ-1-04	1.05	42.00	40.00
AZ-1-11	1.10	35.00	31.82
AZ-1-21	1.08	28.00	25.93
AZ-1-42	1.06	10.00	9.43
AZ-2-00	2.20	85.00	38.64
AZ-2-04	2.16	76.00	35.19
AZ-2-11	2.09	69.00	33.01
AZ-2-21	2.18	52.00	23.85
AZ-2-42	2.22	35.00	15.76
UI-1-00	0.95	28.00	29.47
UI-1-11	1.01	22.00	21.78
UI-1-21	1.04	19.00	18.27
UI-1-42	0.94	10.00	10.64

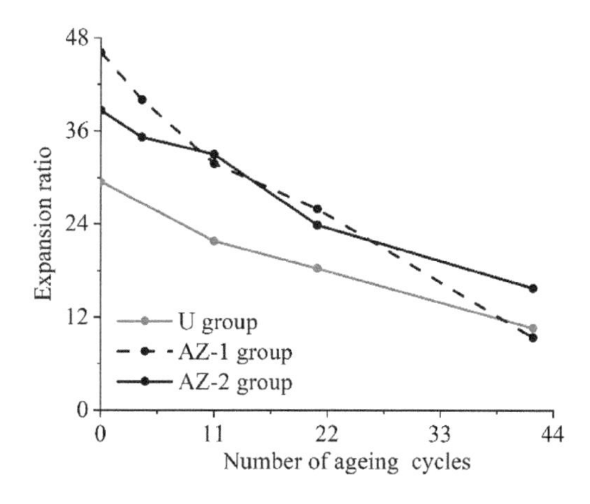

Figure 5.11 Reduction of expansion ratio with number of cycles of accelerated ageing test.

5.3 ASSESSMENT OF INSULATIVE PROPERTIES OF AGED INTUMESCENT COATINGS

5.3.1 Effect on the temperature elevation of steel substrates

The average steel temperature–time relationships obtained from the tests presented in the Section 5.2 are compared in Figure 5.12 to show the effect of the ageing of an intumescent coating on the temperature elevation of the steel substrates exposed to fire. It can be seen that, compared to specimens

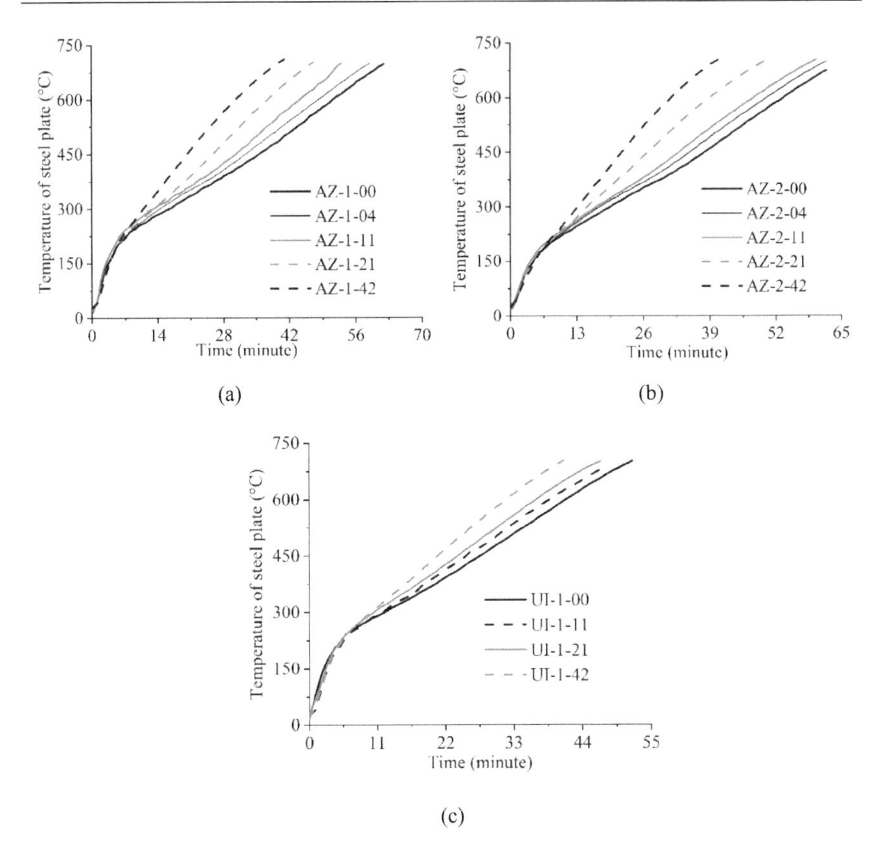

Figure 5.12 Effect of the ageing of a coating on protected steel substrate temperature–time relationship. (a) Type-A specimens with 1 mm coating. (b) Type-A specimens with 2 mm coating. (c) Type-U specimens.

without ageing, there is a sharp increase in the steel substrate temperature after a certain number of cycles of an ageing test on intumescent coatings. For a type-A coating, 11 cycles of an ageing test (representing five years in service) appear to mark the beginning of a sharp increase in the steel substrate temperature. For a type-U coating, the increase in steel temperature appears to be more, over the entire range of ageing cycles.

Figure 5.13 presents the steel substrate temperatures for the specimens which experienced different cycles of the ageing test at the same time as the specimens without ageing reached 400, 500 and 600°C. Furthermore, Table 5.4 lists the fire resistance times that may be achieved by different specimens if the steel limiting temperature is specified at 400, 500, 600 and 700°C respectively.

Obviously, the increase in the steel substrate temperature or the reduction in the fire resistance time, due to ageing which leads to degradation in the insulative properties of intumescent coatings, is significant, and which we need to pay attention to for the whole life safety of steel structures against fire.

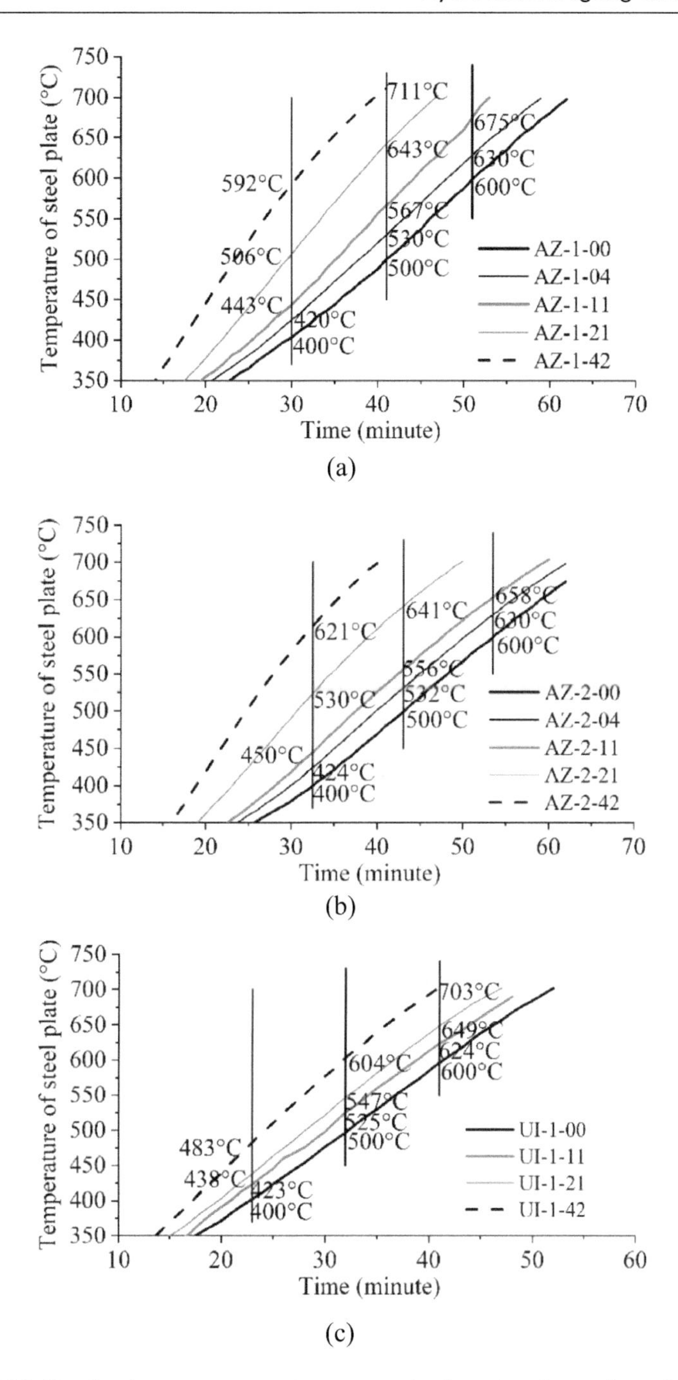

Figure 5.13 Steel substrate temperatures reached at the time when the specimens without ageing reached 400/500/600°C. (a) Type-A specimens with 1 mm coating. (b) Type-A specimens with 2 mm coating. (c) Type-U specimens.

Table 5.4 Fire resistance times of specimens (minutes)

	Limiting steel temperature (°C)			
Specimen	*400*	*500*	*600*	*700*
AZ-1-00	30	41	51	63
AZ-1-04	27	38	48	59
AZ-1-11	25	35	44	53
AZ-1-21	22	30	37	47
AZ-1-42	17	23	30	40
AZ-2-00	33	43	54	66
AZ-2-04	30	40	50	62
AZ-2-11	28	37	48	59
AZ-2-21	23	31	39	50
AZ-2-42	19	25	31	40
UI-1-00	23	32	41	52
UI-1-11	21	30	39	49
UI-1-21	20	28	37	47
UI-1-42	17	24	32	41

5.3.2 Effect on the effective thermal conductivity of coatings

The effective thermal conductivity specified in Chapter 2 may be used to indicate the overall effects of hydrothermal ageing on the fire performance of intumescent coatings (CEN 2013). The variation of the average effective thermal conductivity of these coatings with the number of ageing cycles within the range of steel substrate temperatures from 500 to 600°C is presented in Figure 5.14.

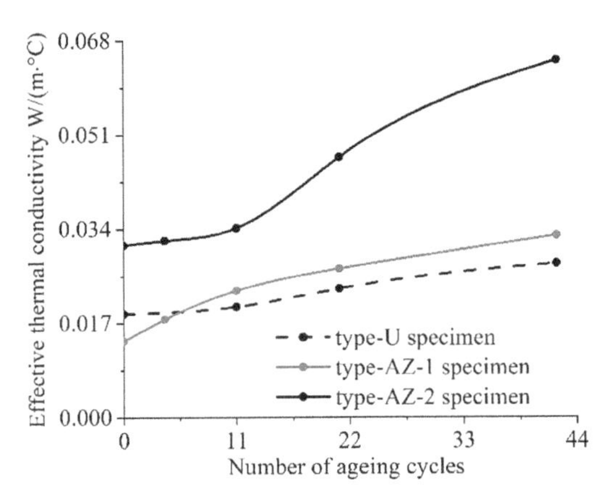

Figure 5.14 Effect of ageing on the effective thermal conductivity of intumescent coatings.

It can be seen from Figure 5.14 that the ageing effect on the insulative property of intumescent coatings is not only relevant to the cycles of ageing, but also to the type and thickness of coatings. More cycles of ageing lead to a greater increase of effective thermal conductivity or a greater reduction in heat insulation. After 21 and 42 cycles of hydrothermal ageing, the effective thermal conductivity of type-U intumescent coating was 25 and 49% higher respectively than that of the fresh coating, and the effective thermal conductivity of type-A intumescent coating was 94 and 138% higher respectively. The thicker 2 mm coating delays the process slightly so that the large change occurs between 11 and 21 cycles of hydrothermal ageing instead of after 4–11 cycles for the thinner 1 mm coating.

5.4 SUMMARY

This chapter has presented the results of a comprehensive experimental study to provide some quantitative information on the degraded fire protection performance of intumescent coatings to steel structures due to ageing. Accelerated environmental changing was employed to simulate the ageing effect on these coatings. The aged coatings were subjected to XPS and FTIR tests to measure their change of element contents and migration of components in the intumescent coating system, which helps to explain the ageing mechanism. It was found that the ageing process did not cause the chemical components of coatings to change, but the optimum matching of these components in the examined intumescent coatings changed due to migration of the hydrophilic components to the surface of the coating when exposed to the hydrothermal ageing environment. This damaged the expanding ability of the intumescent coatings.

The fire tests on aged intumescent coating specimens revealed some effects of ageing on the insulative performance of these coatings. The expansion ratio of coatings may be reduced by over 70% and the steel substrate temperature may be increased by about 200°C above that of the same substrate specimen with a fresh intumescent coating at the same time exposed to fire. Certainly, more cycles of ageing may lead to a greater increase in effective thermal conductivity or a greater reduction in heat insulation on intumescent coatings. A thicker coating delays the ageing process.

REFERENCES

CEN (European Committee for Standardization). 2013. *Test Methods for Determining the Contribution to the Fire Resistance of Structural Members-Part 8: Applied Reactive Protection to Steel Members*. BS EN 13381-8: 2013. Brussels: CEN.

EOTA (European Organization for Technical Assessment). 2017. *European Assessment Document-Fire Protection Products, Reactive Coatings for Fire Protection of Steel Elements*. EAD 350402-00-1106: 2017. Brussels: EOTA.

ISO (International Organization for Standardization). 1975. *Fire Resistance Tests-Elements of Building Construction*. ISO 834: 1975. Geneva: ISO.

Jimenez, Maude, Duquesne, Sophie, and Bourbigot, Serge. 2006. "Intumescent Fire Protective Coating: Toward a Better Understanding of Their Mechanism of Action." *Thermochimica Acta* 449(1–2): 16–26.

Mamleev, Vadim Sh., Bekturov, Esen A., and Gibolv, Konstantin M. 1998. "Dynamics of Intumescence of Fire-Retardant Polymeric Materials." *Journal of Applied Polymer Science* 70(8): 1523–1542.

Wang, Ling-Ling. 2011. "The Ageing of Intumescent Coating and Its Effect on Fire Resistance Behaviour of Steel Structure." PhD diss., Tongji University.

Wang, Ling-Ling, Chen, Bo-Wen, Zhang, Chao, and Li, Guo-Qiang. 2019. "Experimental Study on Insulative Properties of Intumescent Coating Exposed to Standard and Nonstandard Furnace Curves." *Fire and Materials* 43(7): 782–793.

Wang, Ling-Ling, Wang, Yong-Chang, and Li, Guo-Qiang. 2013. "Experimental Study of Hydrothermal Ageing Effects on Insulative Properties of Intumescent Coating for Steel Elements." *Fire Safety Journal* 55: 168–181.

Zhang, Yong, Wang, Yong-Chang, Bailey, Colin G., and Taylor, Andrew P. 2012. "Global Modelling of Fire Protection Performance of Intumescent Coating under Different Cone Calorimeter Heating Conditions." *Fire Safety Journal* 50: 51–62.

Chapter 6

Influence of topcoats on insulation and the anti-ageing performance of intumescent coatings

Intumescent coatings are chemically reactive products and their fire protection properties may degrade after prolonged exposure to environmental conditions. It is important to understand their durability performance so that there is confidence in the safety of intumescent coating protected steel structures throughout their entire life. To improve the durability of these coatings, two generic methods may be used:

1. Add particles into intumescent coatings as modifiers;
2. Apply topcoats.

Applying topcoats is generally adopted in practice to protect intumescent coatings against the ageing effect due to environmental exposure. Therefore, it is important to investigate whether a topcoat is effective in fulfilling this requirement. However, a topcoat may exert a negative influence on the expansion of intumescent coatings and then degrade the fire protection performance of coating chars. This chapter presents the results of two comprehensive experimental studies on the effect of topcoats on fire protection performance and the anti-ageing properties of intumescent coatings.

6.1 EFFECT OF TOPCOATS ON RESTRAINING EXPANSION AND THE THERMAL RESISTANCE OF INTUMESCENT COATINGS

To investigate the effect of topcoats on the fire protection performance of intumescent coatings, an experimental study was conducted by Xu et al. (2018, 28). In Xu et al. (2018, 26), a simple quantity for the constant effective thermal conductivity was used to represent the insulation properties of intumescent coatings for considering the topcoat effect under various conditions, including type and thickness of intumescent coating, the number of topcoat layers, the section factor of the steel specimen and the heating regime.

DOI: 10.1201/9781003287919-6

6.1.1 Specimen preparation

Test specimens were prepared by applying intumescent coatings and topcoats to steel plates and steel elements with other sections. A total of 112 specimens were tested, including 88 steel plate specimens, 12 steel I-section specimens and 12 steel C-section specimens. Two types of commercially available intumescent coatings were applied, to be referred to as type A coating and type B coating respectively. Type A coating is a single component water-borne intumescent coating, containing low volatile organic compounds, and type B coating is a single component solvent-based intumescent coating. Both type A and type B coatings were applied to the steel plate specimens and only type A was applied to steel I-section specimens and C-section specimens. Three nominal values of dry film thickness (DFT) were used for both type A and type B coatings; they are 0.6, 1.4 and 2.2 mm for type A coating and 0.6, 1.2 and 2.0 mm for type B coating, respectively. The different DFTs of coatings were intended to provide 30, 60 and 90 minute ratings of fire resistance for steel elements respectively.

A two-component, high gloss, polyurethane topcoat was applied on top of the intumescent coatings. The topcoat was supplied by the same manufacturer as the intumescent coatings and was compatible with the above two types of intumescent coatings. The topcoat thickness per layer was approximately 35 μm and zero, one, three and five layers of topcoat were applied respectively to identify the effect of topcoat layer numbers on the fire protection behavior of intumescent coatings. Two thermocouples were embedded in each steel plate specimen and three thermocouples were embedded in each steel I-section and C-section specimen, as shown in Figure 6.1. The main parameters of the test specimens are listed in Table 6.1.

6.1.2 Fire tests

The fire tests were conducted in a furnace whose internal dimensions were 1.0 m × 1.0 m × 1.2 m, as shown in Figure 6.2. Four steel plate specimens can be hung on supporting beams inside the furnace for one fire test.

The furnace temperatures were regulated according to the ISO 834 standard (ISO 1975) for simulating a standard fire and according to the code of CECS 200 (CECS 2006a) for simulating large space fire curves. Figure 6.3 illustrates measurements and predictions of temperature–time curves of the furnace.

6.1.3 Test results

6.1.3.1 Appearance and expansion of intumescent coatings

The expanded intumescent coating char of the steel plate and other section specimens with different numbers of topcoat layers are compared in Figures 6.4 and 6.5 respectively. The surface of the intumescent coating char without a topcoat was uniform with small bubbles on it, as shown

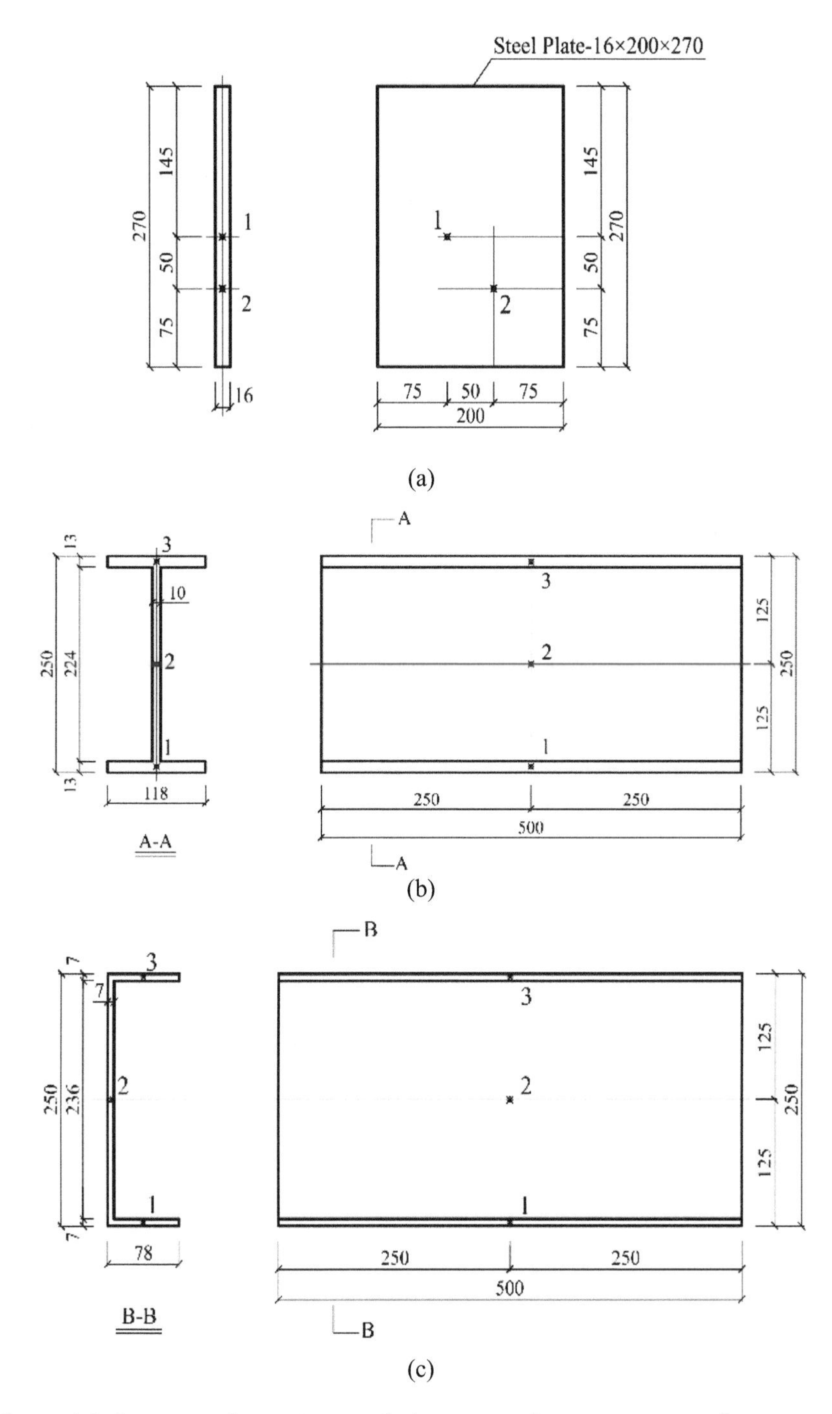

Figure 6.1 Specimen dimensions and thermocouple arrangements for topcoat effect tests. (a) Steel plate. (b) I-section. (c) C-section.

Table 6.1 Main parameters of test specimens

Specimen ID	Steel substrate	Section factor F_i/V (m^{-1})	Coating type	Target DFT (mm)	Topcoat layers	Heating condition
PA06T0F0	Plate	142.4	A	0.6	0	ISO
PA06T0F1	Plate	142.4	A	0.6	1	ISO
PA06T0F3	Plate	142.4	A	0.6	3	ISO
PA06T0F5	Plate	142.4	A	0.6	5	ISO
PA14T0F0	Plate	142.4	A	1.4	0	ISO
PA14T0F1	Plate	142.4	A	1.4	1	ISO
PA14T0F3	Plate	142.4	A	1.4	3	ISO
PA14T0F5	Plate	142.4	A	1.4	5	ISO
PA22T0F0	Plate	142.4	A	2.2	0	ISO
PA22T0F1	Plate	142.4	A	2.2	1	ISO
PA22T0F3	Plate	142.4	A	2.2	3	ISO
PA22T0F5	Plate	142.4	A	2.2	5	ISO
IA06T0F0	I-shaped	177.9	A	0.6	0	ISO
IA06T0F1	I-shaped	177.9	A	0.6	1	ISO
IA06T0F3	I-shaped	177.9	A	0.6	3	ISO
IA06T0F5	I-shaped	177.9	A	0.6	5	ISO
IA14T0F0	I-shaped	177.9	A	1.4	0	ISO
IA14T0F1	I-shaped	177.9	A	1.4	1	ISO
IA14T0F3	I-shaped	177.9	A	1.4	3	ISO
IA14T0F5	I-shaped	177.9	A	1.4	5	ISO
IA22T0F0	I-shaped	177.9	A	2.2	0	ISO
IA22T0F1	I-shaped	177.9	A	2.2	1	ISO
IA22T0F3	I-shaped	177.9	A	2.2	3	ISO
IA22T0F5	I-shaped	177.9	A	2.2	5	ISO
CA06T0F0	C-shaped	228.6	A	0.6	0	ISO
CA06T0F1	C-shaped	228.6	A	0.6	1	ISO
CA06T0F3	C-shaped	228.6	A	0.6	3	ISO
CA06T0F5	C-shaped	228.6	A	0.6	5	ISO
CA14T0F0	C-shaped	228.6	A	1.4	0	ISO
CA14T0F1	C-shaped	228.6	A	1.4	1	ISO
CA14T0F3	C-shaped	228.6	A	1.4	3	ISO
CA14T0F5	C-shaped	228.6	A	1.4	5	ISO
CA22T0F0	C-shaped	228.6	A	2.2	0	ISO
CA22T0F1	C-shaped	228.6	A	2.2	1	ISO
CA22T0F3	C-shaped	228.6	A	2.2	3	ISO
CA22T0F5	C-shaped	228.6	A	2.2	5	ISO
PA06T1F0	Plate	142.4	A	0.6	0	T1
PA06T1F1	Plate	142.4	A	0.6	1	T1
PA06T1F3	Plate	142.4	A	0.6	3	T1
PA06T1F5	Plate	142.4	A	0.6	5	T1

(Continued)

Table 6.1 (Continued) Main parameters of test specimens

Specimen ID	*Steel substrate*	*Section factor F_i/V (m^{-1})*	*Coating type*	*Target DFT (mm)*	*Topcoat layers*	*Heating condition*
PA14T1F0	Plate	142.4	A	1.4	0	T1
PA14T1F1	Plate	142.4	A	1.4	1	T1
PA14T1F3	Plate	142.4	A	1.4	3	T1
PA14T1F5	Plate	142.4	A	1.4	5	T1
PA22T1F0	Plate	142.4	A	2.2	0	T1
PA22T1F1	Plate	142.4	A	2.2	1	T1
PA22T1F3	Plate	142.4	A	2.2	3	T1
PA22T1F5	Plate	142.4	A	2.2	5	T1
PA06T2F0	Plate	142.4	A	0.6	0	T2
PA06T2F1	Plate	142.4	A	0.6	1	T2
PA06T2F3	Plate	142.4	A	0.6	3	T2
PA06T2F5	Plate	142.4	A	0.6	5	T2
PA14T2F0	Plate	142.4	A	1.4	0	T2
PA14T2F1	Plate	142.4	A	1.4	1	T2
PA14T2F3	Plate	142.4	A	1.4	3	T2
PA14T2F5	Plate	142.4	A	1.4	5	T2
PA22T2F0	Plate	142.4	A	2.2	0	T2
PA22T2F1	Plate	142.4	A	2.2	1	T2
PA22T2F3	Plate	142.4	A	2.2	3	T2
PA22T2F5	Plate	142.4	A	2.2	5	T2
PB06T0F0	Plate	142.4	B	0.6	0	ISO
PB06T0F1	Plate	142.4	B	0.6	1	ISO
PB06T0F3	Plate	142.4	B	0.6	3	ISO
PB06T0F5	Plate	142.4	B	0.6	5	ISO
PB12T0F0	Plate	142.4	B	1.2	0	ISO
PB12T0F1	Plate	142.4	B	1.2	1	ISO
PB12T0F3	Plate	142.4	B	1.2	3	ISO
PB12T0F5	Plate	142.4	B	1.2	5	ISO
PB20T0F0	Plate	142.4	B	2.0	0	ISO
PB20T0F1	Plate	142.4	B	2.0	1	ISO
PB20T0F3	Plate	142.4	B	2.0	3	ISO
PB20T0F5	Plate	142.4	B	2.0	5	ISO

in Figures 6.4(a) and 6.5(a). When a topcoat was applied, the surface of the char was that of the topcoat which was hard with big bubbles and cracks. The topcoat started to burn when the steel temperature was about 200°C, and the flame was extinguished after the combustible part of the topcoat was consumed when the steel temperature reached approximately 350°C. The big bubbles in the topcoat layer were shaped by the gases released from the intumescent coating during expansion, which coalesced under the topcoat. The thicker the topcoat was, the harder the surface and the bigger the bubbles formed.

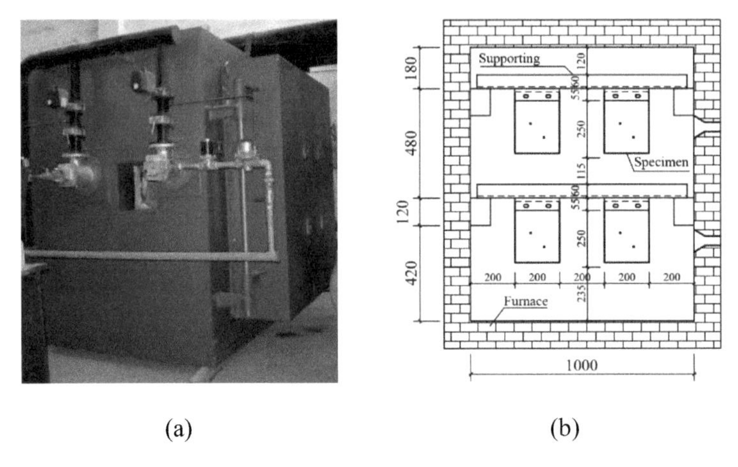

Figure 6.2 Furnace and its internal dimensions. (a) Furnace. (b) Internal dimensions of the furnace.

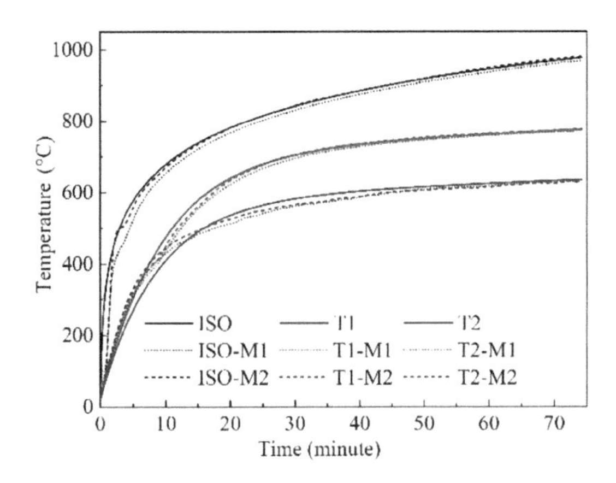

Figure 6.3 Comparison of fire curves.

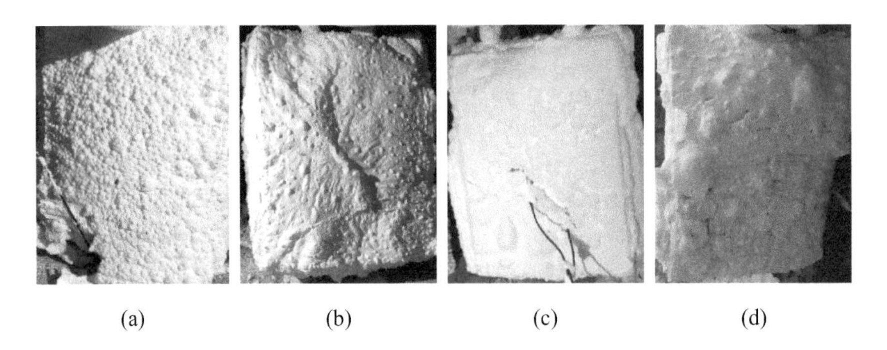

Figure 6.4 Expanded intumescent coating char of steel plate specimens after exposure to the ISO 834 standard fire. (a) PA22T0F0. (b) PA22T0F1. (c) PA22T0F3. (d) PA22T0F5.

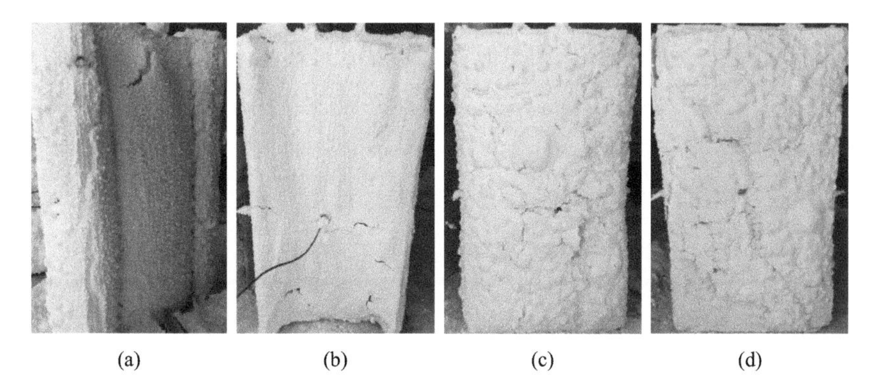

(a) (b) (c) (d)

Figure 6.5 Expanded intumescent coating char of steel I-section and C-section specimens after exposure to the ISO 834 standard fire. (a) IA14T0F0. (b) IA06T0F1. (c) CA06T0F3. (d) CA06T0F5.

Tables 6.2 and 6.3 list the expansion ratios of intumescent coatings after exposure to the ISO 834 standard fire and large space fires respectively. The final thickness of the intumescent char was measured over a relatively uniform area on the surface, and the average measurements of five locations were taken as the representative value.

It can be seen from Tables 6.2 and 6.3 that applying a topcoat has little effect on the expansion ratios of the waterborne coating exposed to the ISO standard fire. However, applying a topcoat was found to have a detrimental effect on the expansion ratios of the waterborne coating when exposed to the large space fires, which is due to the temperature of these fires being much lower and the intumescent coatings having not fully expanded (Xu et al. 2018, 34). As regards to the solvent-based coating, the expansion ratios of the coating with the topcoat were significantly reduced, regardless as to whether the fire type is post-flashover or large space.

6.1.3.2 Steel temperatures

Some results of the average steel temperatures are compared in Figure 6.6 for specimens with different numbers of topcoat layers and exposed to the ISO 834 standard fire. It can be seen that for the specimens with waterborne intumescent coating, the steel temperatures clustered within a relatively close band, indicating that the topcoat had little effect on the protected steel temperature. In contrast, for the specimens with a solvent-based intumescent coating, the steel temperature was noticeably higher when a topcoat was used, which is in accordance with the decease of expansion ratio, leading to the reduction of insulation due to a topcoat.

Steel temperatures are presented in Figure 6.7 for specimens protected by a waterborne intumescent coating with different numbers of topcoat layers under the large space fire T2. A notable protected steel substrate temperature increase can be observed when using a topcoat, which concurs with the reductions in intumescent coating expansion, as shown in Table 6.3.

Table 6.2 Expansion ratios of intumescent coatings after exposure to the ISO standard fire

Specimen	*Initial thickness (mm)*	*Final thickness (mm)*	*Expansion ratio*
PA06T0F0	0.539	15.50	28.8
PA06T0F1	0.542	19.15	35.3
PA06T0F3	0.560	11.25	20.1
PA06T0F5	0.533	16.80	31.5
PA14T0F0	1.456	28.97	19.9
PA14T0F1	1.432	36.23	25.3
PA14T0F3	1.388	38.60	27.8
PA14T0F5	1.320	30.35	23.0
PA22T0F0	2.274	50.94	22.4
PA22T0F1	2.141	50.95	23.8
PA22T0F3	2.146	50.65	23.6
PA22T0F5	2.122	45.63	21.5
IA06T0F0	0.553	18.00	32.5
IA06T0F1	0.717	25.00	34.9
IA06T0F3	0.830	25.40	30.6
IA06T0F5	0.719	24.60	34.2
IA14T0F0	1.196	55.20	46.2
IA14T0F1	1.228	60.30	49.1
IA14T0F3	1.319	55.50	42.1
IA14T0F5	1.234	54.90	44.5
IA22T0F0	2.126	60.30	28.4
IA22T0F1	2.107	60.50	28.7
IA22T0F3	2.236	73.00	32.6
IA22T0F5	2.320	70.70	30.5
CA06T0F0	0.574	15.80	27.5
CA06T0F1	0.625	25.00	40.0
CA06T0F3	0.706	22.50	31.9
CA06T0F5	0.724	20.00	27.6
CA14T0F0	1.343	35.90	26.7
CA14T0F1	1.294	50.10	38.7
CA14T0F3	1.269	49.90	39.3
CA14T0F5	1.199	51.00	42.5
CA22T0F0	2.048	74.00	36.1
CA22T0F1	2.370	75.00	31.6
CA22T0F3	2.673	74.90	28.0
CA22T0F5	2.436	75.30	30.9
PB06T0F0	0.643	15.10	23.5
PB06T0F1	0.628	12.00	19.1
PB06T0F3	0.652	15.85	24.3

(Continued)

Table 6.2 (Continued) Expansion ratios of intumescent coatings after exposure to the ISO standard fire

Specimen	*Initial thickness (mm)*	*Final thickness (mm)*	*Expansion ratio*
PB06T0F5	0.632	9.80	15.5
PB12T0F0	1.197	24.17	20.2
PB12T0F1	1.287	22.65	17.6
PB12T0F3	1.166	20.75	17.8
PB12T0F5	1.225	18.75	15.3
PB20T0F0	1.951	40.00	20.5
PB20T0F1	2.015	40.90	20.3
PB20T0F3	2.059	27.80	13.5
PB20T0F5	2.005	39.90	19.9

Table 6.3 Expansion ratios of specimens after exposure to large space fires

Specimen	*Initial thickness (mm)*	*Final thickness (mm)*	*Expansion ratio*
PA06T1F0	0.641	10.00	15.6
PA06T1F1	0.476	3.00	6.3
PA06T1F3	0.529	5.00	9.5
PA06T1F5	0.563	5.00	8.9
PA14T1F0	1.403	35.00	25.0
PA14T1F1	1.370	32.50	23.7
PA14T1F3	1.512	35.00	23.2
PA14T1F5	1.534	40.50	26.4
PA22T1F0	2.267	60.30	26.6
PA22T1F1	2.206	45.10	20.4
PA22T1F3	2.153	30.30	14.1
PA22T1F5	2.253	40.00	17.8
PA06T2F0	0.576	5.00	8.7
PA06T2F1	0.477	3.00	6.3
PA06T2F3	0.546	3.00	5.5
PA06T2F5	0.559	3.00	5.4
PA14T2F0	1.512	32.50	20.0
PA14T2F1	1.243	22.50	18.1
PA14T2F3	1.386	27.50	19.8
PA14T2F5	1.424	17.50	12.3
PA22T2F0	2.144	37.50	17.3
PA22T2F1	2.258	37.50	16.6
PA22T2F3	2.323	41.00	17.7
PA22T2F5	2.281	40.00	17.5

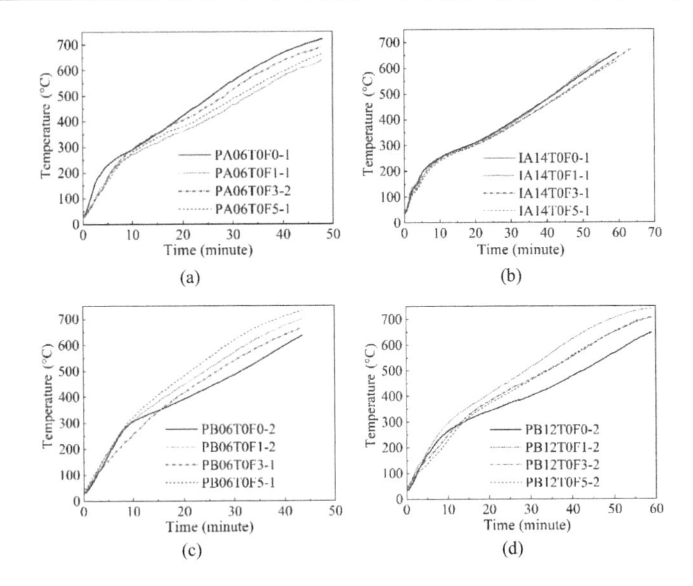

Figure 6.6 Comparison of steel temperatures for specimens protected by both types of intumescent coating with different numbers of topcoat layers under the ISO 834 standard fire. (a) IA 06. (b) IA 14. (c) PB 06. (d) PB 12.

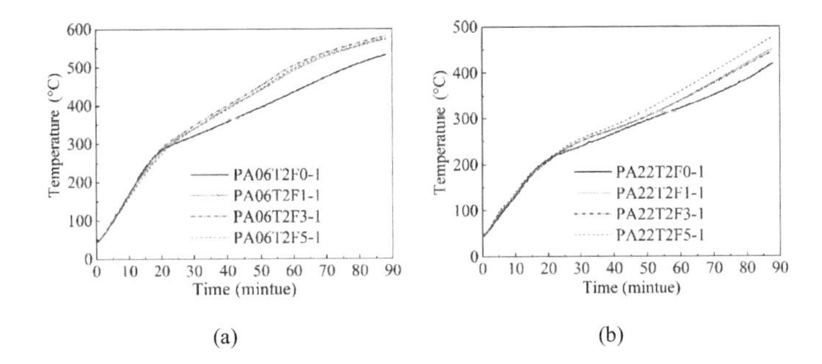

Figure 6.7 Comparison of steel temperatures for specimens protected by a waterborne intumescent coating with different numbers of topcoat layers under the large space fire T2. (a) PA 06. (b) PA 22.

6.1.3.3 Constant effective thermal conductivity

The constant effective thermal conductivity, λ_e, specified in Chapter 2, can be used as an indicator for the comprehensive insulation of intumescent coatings (Li et al. 2016). The constant effective thermal conductivities of the test specimens exposed to the ISO 834 standard fire are listed in Table 6.4. Base on the results in Table 6.4, the average variation of the constant effective thermal conductivity of the waterborne intumescent coatings and solvent-based intumescent coatings with a topcoat are compared to that without a topcoat, as presented in Tables 6.5 and 6.6 respectively. It can be seen that a

Table 6.4 Constant effective thermal conductivities of intumescent coatings with and without topcoats

Specimen	*Number of topcoat layers*	*Target DFT of intumescent coating (mm)*	*Substrate type*	λ_e *W/m/k*
PA06T0F0	0	0.6	Plate	0.013522
PA06T0F1	1	0.6	Plate	0.011631
PA06T0F3	3	0.6	Plate	0.01423
PA06T0F5	5	0.6	Plate	0.010555
PA14T0F0	0	1.4	Plate	0.02255
PA14T0F1	1	1.4	Plate	0.023115
PA14T0F3	3	1.4	Plate	0.021059
PA14T0F5	5	1.4	Plate	0.019182
PA22T0F0	0	2.2	Plate	0.027947
PA22T0F1	1	2.2	Plate	0.033222
PA22T0F3	3	2.2	Plate	0.032398
PA22T0F5	5	2.2	Plate	0.030611
IA06T0F0	0	0.6	I-section	0.01046
IA06T0F1	1	0.6	I-section	0.009983
IA06T0F3	3	0.6	I-section	0.010618
IA06T0F5	5	0.6	I-section	0.010682
IA14T0F0	0	1.4	I-section	0.013596
IA14T0F1	1	1.4	I-section	0.015135
IA14T0F3	3	1.4	I-section	0.014069
IA14T0F5	5	1.4	I-section	0.012765
IA22T0F0	0	2.2	I-section	0.023186
IA22T0F1	1	2.2	I-section	0.021483
IA22T0F3	3	2.2	I-section	0.020161
IA22T0F5	5	2.2	I-section	0.023079
CA06T0F0	0	0.6	C-section	0.010089
CA06T0F1	1	0.6	C-section	0.008728
CA06T0F3	3	0.6	C-section	0.009702
CA06T0F5	5	0.6	C-section	0.011213
CA14T0F0	0	1.4	C-section	0.01318
CA14T0F1	1	1.4	C-section	0.013957
CA14T0F3	3	1.4	C-section	0.012332
CA14T0F5	5	1.4	C-section	0.012138
CA22T0F0	0	2.2	C-section	0.023829
CA22T0F1	1	2.2	C-section	0.02494
CA22T0F3	3	2.2	C-section	0.024688
CA22T0F5	5	2.2	C-section	0.02352
PB06T0F0	0	0.6	Plate	0.013519
PB06T0F1	1	0.6	Plate	0.016474

(Continued)

Table 6.4 (Continued) Constant effective thermal conductivities of intumescent coatings with and without topcoats

Specimen	*Number of topcoat layers*	*Target DFT of intumescent coating (mm)*	*Substrate type*	*λ_e W/m/k*
PB06T0F3	3	0.6	Plate	0.016066
PB06T0F5	5	0.6	Plate	0.018654
PB12T0F0	0	1.2	Plate	0.020537
PB12T0F1	1	1.2	Plate	0.025403
PB12T0F3	3	1.2	Plate	0.025611
PB12T0F5	5	1.2	Plate	0.025144
PB20T0F0	0	2.0	Plate	0.025831
PB20T0F1	1	2.0	Plate	0.029775
PB20T0F3	3	2.0	Plate	0.02775
PB20T0F5	5	2.0	Plate	0.028763

Table 6.5 Effect of topcoats on constant effective thermal conductivity of a waterborne intumescent coating

Number of topcoat layers	*Section type*	*Difference ratio (%) Coatings with different nominal DFT*			*Average difference ratio (%)*
		0.6 (mm)	*1.4 (mm)*	*2.2 (mm)*	
	Plate	−14.00	2.50	18.90	
1	I-section	−4.60	11.30	−7.30	0.43
	C-section	−13.50	5.90	4.70	
	Plate	5.20	−6.60	15.90	
3	I-section	1.50	3.50	−13.00	−0.01
	C-section	−3.80	−6.40	3.60	
	Plate	−21.90	−14.90	9.50	
5	I-section	2.10	−6.10	−0.50	−2.50
	C-section	11.10	−7.90	6.10	

Table 6.6 Effect of topcoats on constant effective thermal conductivity of a solvent-based intumescent coating

Number of topcoat layers	*Difference with no topcoat (0.6 DFT) (%)*	*Difference with no topcoat (1.2 DFT) (%)*	*Difference with no topcoat (2.0 DFT) (%)*	*Average difference ratio (%)*
1	21.90	23.70	15.30	20.30
3	18.80	24.70	7.40	16.97
5	38.00	22.40	11.40	23.93

topcoat has little effect on the constant effective thermal conductivity of the waterborne intumescent coatings, while having a significant effect on that of a solvent-based intumescent coating.

6.2 EFFECT OF TOPCOATS ON THE AGEING OF INTUMESCENT COATINGS

A comprehensive experimental study was conducted by Wang et al. (2020) to investigate the effects of two types of topcoat on the fire protection performance of intumescent coatings under accelerated ageing. In the study, after subjecting intumescent-coating-protected steel plates to different cycles of accelerated ageing, the specimens were tested in fire.

6.2.1 Specimen preparation

The test specimens were prepared by applying an intumescent coating and topcoats to steel plates. A total of 105 specimens were tested, 60 of which were protected by the intumescent coating with a topcoat of epoxy polyurethane (to be referred to as type-P specimens), 30 by the intumescent coating with a topcoat of aliphatic acrylic polyurethane (to be referred to as type-S specimens) and 15 with the intumescent coating but no topcoat (to be referred to as type-N specimens). In all specimens, a solvent-based intumescent coating was applied and whose main ingredients were ammonium polyphosphate (APP—pentaerythritol (PER)—melamine (MEL) in a single component acrylic resin. The nominal DFT of the intumescent coating was 1 mm. The actual intumescent coating DFT of each specimen was taken as the average value of measurements at 16 locations, varying in the range of 0.95 to 1.09 mm. The topcoat thickness per layer was about 25 μm. The 60 type-P specimens consisted of three replicates of four different topcoat layers (1, 2, 3 and 4) and five different cycles of ageing (0, 21, 42, 63 and 84 cycles simulating 0, 10, 20, 30 and 40 years in service respectively). The 30 type-S specimens were three replicates of five ageing cycles and two different topcoat layers (1 and 2).

The substrate steel plate of the test specimens was measured at 200 mm by 270 mm by 16 mm thick, giving a steel section factor of 142 m^{-1}, typical of those in practice. Three thermocouples (2.0 mm diameter, type K) were embedded in each steel plate in pre-drilled holes to half the thickness of the steel plate, as shown in Figure 6.8, where d_i is the intumescent coating DFT and d_t is the topcoat DFT. Table 6.7 lists the main parameters of the test specimens.

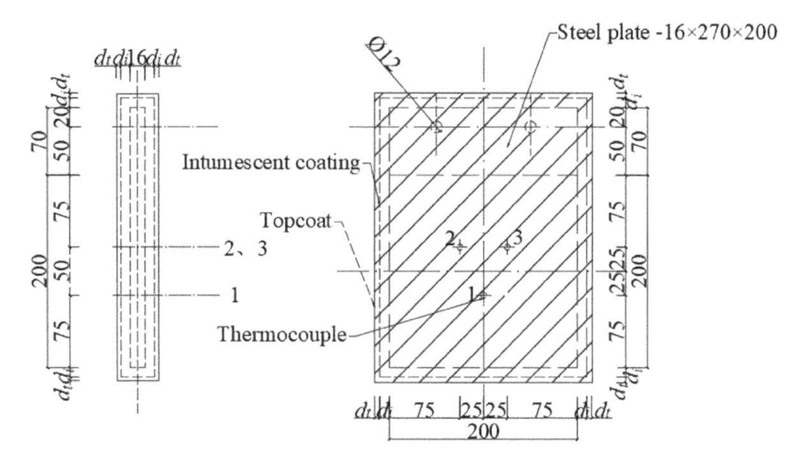

Figure 6.8 Specimen dimensions.

6.2.2 Hydrothermal ageing test

The accelerated hydrothermal ageing test was performed on the specimens according to European guideline EAD 350402-00-1106 (EOTA 2017) and the conditions of an indoor environment with above zero temperature and high humidity was adopted (Wang et al. 2020), which is specified in Section 5.2.2.

The surface appearances of the specimens are shown in Figures 6.9–6.11 after different cycles of accelerated ageing tests. It can be seen that for intumescent coating specimens without a topcoat, bumps appeared on the surface after 21 cycles of the hydrothermal ageing test, as shown in Figure 6.9(b). Wrinkles can be clearly seen in Figure 6.9(c, d) after 42 and 63 cycles of the ageing test. After 84 cycles of the ageing test, the intumescent coating surface was uneven, with large bumps, as shown in Figure 6.9(e). However, when a topcoat was applied to the specimen, phenomena such as bumps, an uneven surface or wrinkles were delayed or reduced. For example, the surface appearance of specimen P-2-63 with an epoxy polyurethane topcoat and 63 cycles of the ageing test, as shown in Figure 6.10(b), is similar to that of specimen N-0-21 with no topcoat and 21 cycles of the ageing test, as shown in Figure 6.9(b). Furthermore, type-S specimens with an aliphatic acrylic polyurethane topcoat shown in Figure 6.11 did not suffer any noticeable change in appearance even after 84 cycles of the ageing test.

6.2.3 Fire test

After the accelerated ageing test, four specimens were placed in a furnace and exposed to fire, as in the previous test presented in Section 5.2.3. The furnace temperature was controlled according to the ISO 834 standard

Table 6.7 Main parameters of the test specimens

Topcoat type	*Topcoat layers*	*No. of cycles of accelerated ageing*	*Simulating time in service*	*Specimen ID*
None	0	0	0	N-0-00-*i* (*i*=1-3)
		21	10	N-0-21-*i*(*i*=1-3)
		42	20	N-0-42-*i*(*i*=1-3)
		63	30	N-0-63-*i*(*i*=1-3)
		84	40	N-0-84-*i*(*i*=1-3)
Thermoplastic	1	0	0	P-1-00-*i*(*i*=1-3)
		21	10	P-1-21-*i*(*i*=1-3)
		42	20	P-1-42-*i*(*i*=1-3)
		63	30	P-1-63-*i*(*i*=1-3)
		84	40	P-1-84-*i*(*i*=1-3)
Thermoplastic	2	0	0	P-2-00-*i*(*i*=1-3)
		21	10	P-2-21-*i*(*i*=1-3)
		42	20	P-2-42-*i*(*i*=1-3)
		63	30	P-2-63-*i*(*i*=1-3)
		84	40	P-2-84-*i*(*i*=1-3)
Thermoplastic	3	0	0	P-3-00-*i*(*i*=1-3)
		21	10	P-3-21-*i*(*i*=1-3)
		42	20	P-3-42-*i*(*i*=1-3)
		63	30	P-3-63-*i*(*i*=1-3)
		84	40	P-3-84-*i*(*i*=1-3)
Thermoplastic	4	0	0	P-4-00-*i*(*i*=1-3)
		21	10	P-4-21-*i*(*i*=1-3)
		42	20	P-4-42-*i*(*i*=1-3)
		63	30	P-4-63-*i*(*i*=1-3)
		84	40	P-4-84-*i*(*i*=1-3)
Thermosetting	1	0	0	S-1-00-*i*(*i*=1-3)
		21	10	S-1-21-*i*(*i*=1-3)
		42	20	S-1-42-*i*(*i*=1-3)
		63	30	S-1-63-*i*(*i*=1-3)
		84	40	S-1-84-*i*(*i*=1-3)
Thermosetting	2	0	0	S-2-00-*i*(*i*=1-3)
		21	10	S-2-21-*i*(*i*=1-3)
		42	20	S-2-42-*i*(*i*=1-3)
		63	30	S-2-63-*i*(*i*=1-3)
		84	40	S-2-84-*i*(*i*=1-3)

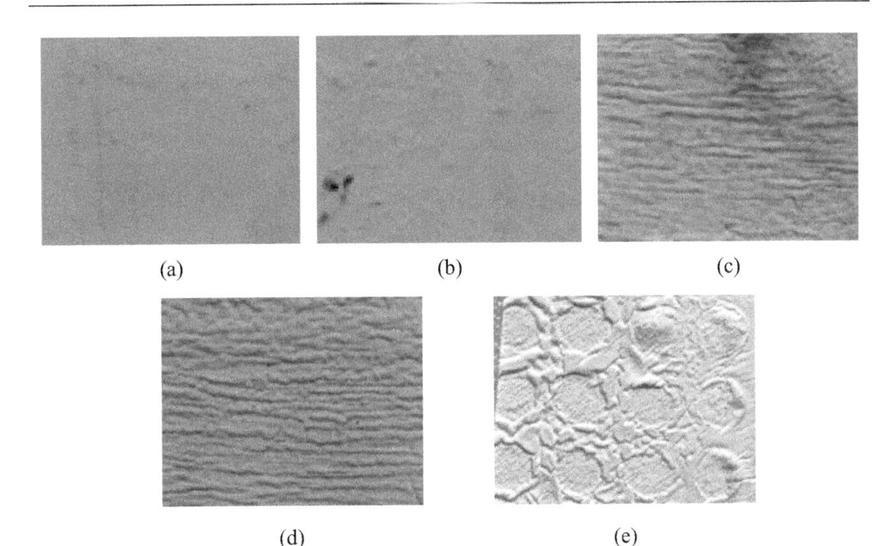

Figure 6.9 Surface appearances of type-N (no topcoat) specimens after different cycles of the hydrothermal ageing test. (a) N-0-00. (b) N-0-21. (c) N-0-42. (d) N-0-63. (e) N-0-84.

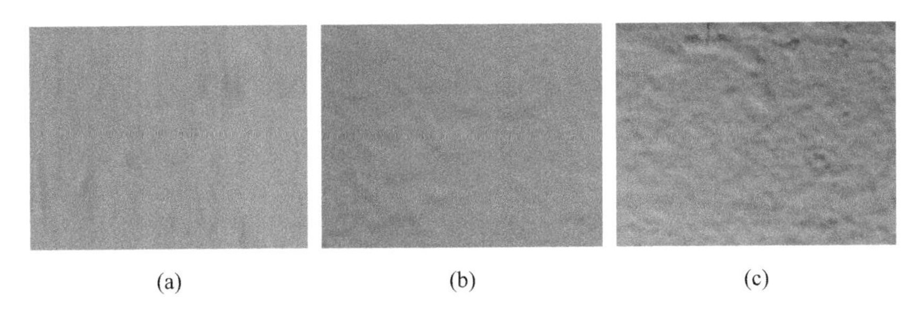

Figure 6.10 Surface appearances of type-P specimens with two layers of topcoats after different cycles of the hydrothermal ageing test. (a) P-2-42. (b) P-2-63. (c) P-2-84.

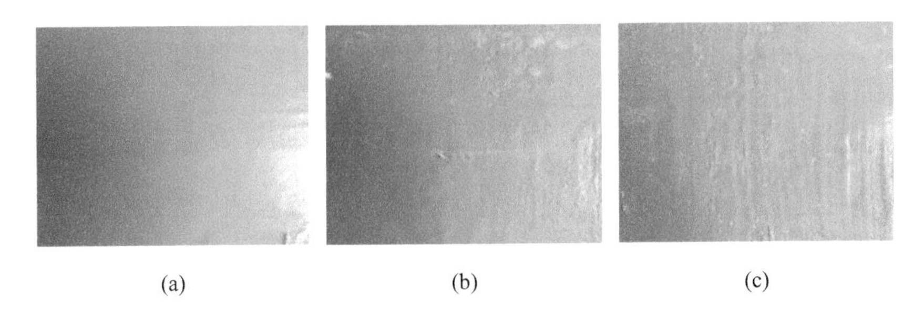

Figure 6.11 Surface appearances of type-S specimens with two layers of topcoats after different cycles of the hydrothermal ageing test. (a) S-2-42. (b) S-2-63. (c) S-2-84.

temperature–time relationship (ISO 1975). The gas temperatures near the specimen surface and the steel temperatures were measured and recorded every 30 seconds continuously. The changing appearance of the specimens were observed from an observation hole placed in the furnace door during fire tests. Each fire test was continued until the steel temperature of the specimens reached 700°C.

6.2.3.1 Surface appearance of specimens

With increasing temperature, the intumescent coatings of all the specimens underwent melting, expansion and forming carbonaceous products which solidified into a multi-cellular char, and char degradation, in the generic sequence of events as observed by others (Shuklin et al. 2004, 200; Morys et al. 2017, 1575). These reactions caused the coatings to swell to form a porous flame-resistant char.

The bubble appearances of different intumescent coating specimens are shown in Figure 6.12 during the expansion and solidifying steps. For specimens without a topcoat, many bubbles appeared on the surface of the

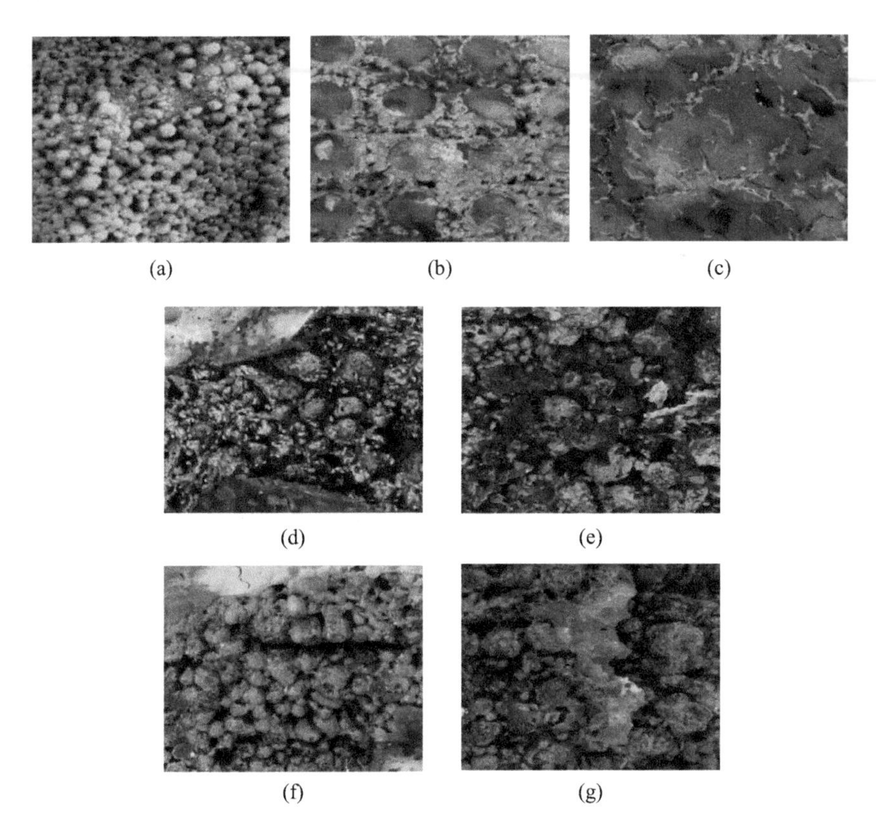

Figure 6.12 Bubble appearance on the surfaces of different specimens. (a) N-0-00. (b) N-0-42. (c) N-0-84. (d) P-1-42. (e) P-1-84. (f) S-1-42. (g) S-1-84.

unaged intumescent coating char, as shown in Figure 6.12(a). The bubbles indicate the effective expansion of intumescent coatings to form a porous char for the fire protection of the steel substrate. After hydrothermal ageing, the number of bubbles decreased drastically due to migration and dissolution of the hydrophilic components in the intumescent coatings (Wang et al. 2006; Wang et al. 2013), more severely for heavier ageing, as shown in Figure 6.12(b, c) for 42 and 84 cycles of ageing tests respectively. In contrast, application of a topcoat delayed or limited the migration of hydrophilic components in intumescent coatings, allowing the chemical reactions of these coatings to remain as designed, as shown in Figures 6.12(d–g) for the specimens with a topcoat. It can be seen that there were still large numbers of bubbles on the surfaces of specimens with a topcoat even after 42 and 84 cycles of ageing tests.

6.2.3.2 Microstructures of intumescent coating chars

The microstructures of the coating chars of the specimens after fire tests were obtained with a scanning electron microscope (SEM) and used for assessing the effect of a topcoat on the ageing of an intumescent coating, as shown in Figure 6.13. The appearance illustrated in Figures 6.13(a–c) indicates that the integrity and consistency of the intumescent coating chars for specimens without a topcoat were severely deteriorated after being exposed to hydrothermal ageing. However, when a topcoat was applied, the integrity and consistence of the intumescent coating chars were much better and pore distributions were much more uniform, as shown in Figures 6.13(d–f).

Figure 6.13 SEM micrographs of intumescent coating chars for different specimens. (a) N-0-00. (b) N-0-42. (c) N-0-84. (d) S-1-00. (e) S-1-42. (f) S-1-84.

6.2.3.3 Effect on expansion ratios

The effect of a topcoat on the expansion ratios of intumescent coatings with ageing was investigated, as shown in Figure 6.14. The expansion ratios of the specimens without a topcoat and without ageing or with only mild ageing were greater than those with a topcoat due to the topcoat restraining effect on the expansion of intumescent coatings. However, for specimens

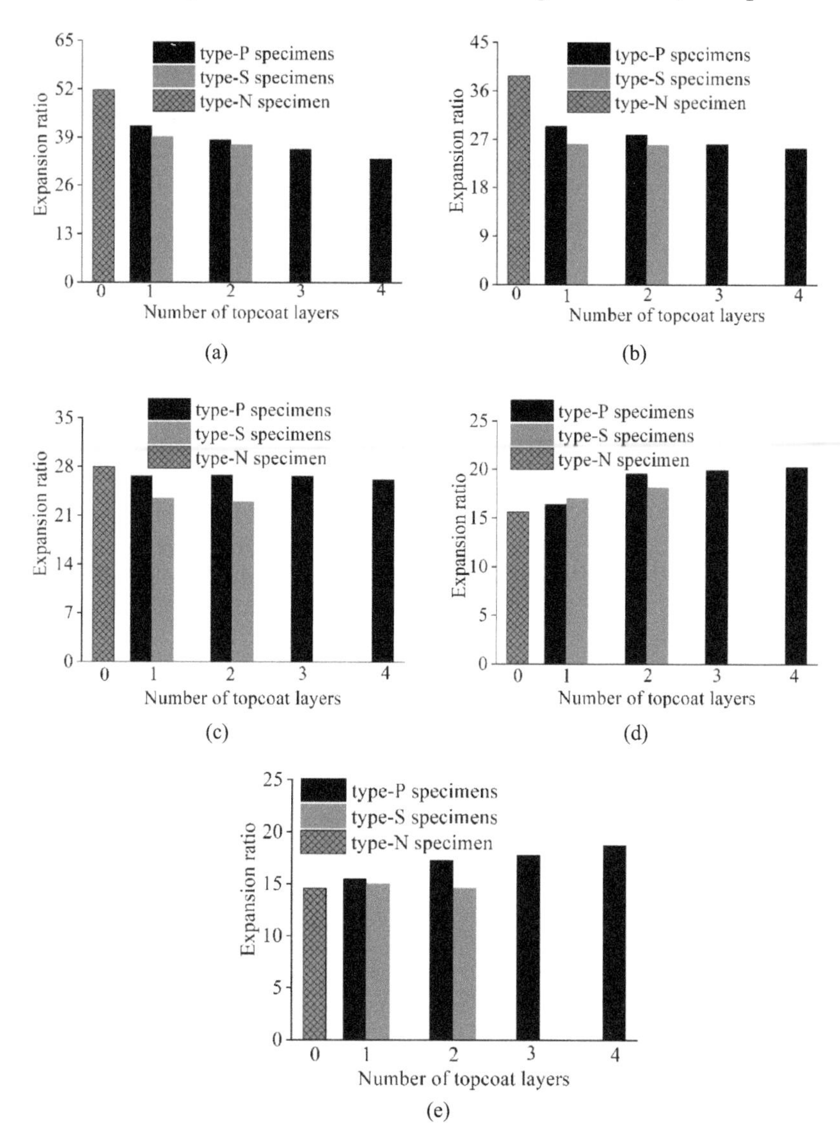

Figure 6.14 Comparison of expansion ratios between specimens with different numbers of topcoat layers after different cycles of ageing. (a) 0 cycles of ageing. (b) 21 cycles of ageing. (c) 42 cycles of ageing. (d) 63 cycles of ageing. (e) 84 cycles of ageing.

with severe ageing (after 63 and 84 cycles of ageing tests), the expansion ratios without a topcoat were less than those with a topcoat because of a reduction of the effective ageing by a topcoat, indicating the beneficial effect of a topcoat on the ageing of intumescent coatings.

6.2.3.4 Effect on the temperature elevations of steel substrates

The variations of the steel temperature elevation of the specimens without and with a topcoat during the fire tests are illustrated in Figures 6.15–6.17. Obviously, for the specimens without a topcoat, the increase in steel temperature due to the degradation in intumescent coatings is significant, as shown in Figure 6.15. This temperature increase is much lower when a topcoat was applied and the ageing effect was more effectively mitigated by increasing the number of topcoat layers, as shown in Figures 6.16 and 6.17.

Steel temperatures at 30 minutes of the test specimens are compared in Figure 6.18 between different numbers of layers of topcoats for different cycles of ageing. They reveal the two opposing effects of topcoats on long-term intumescent coating performance. With no or short duration of ageing,

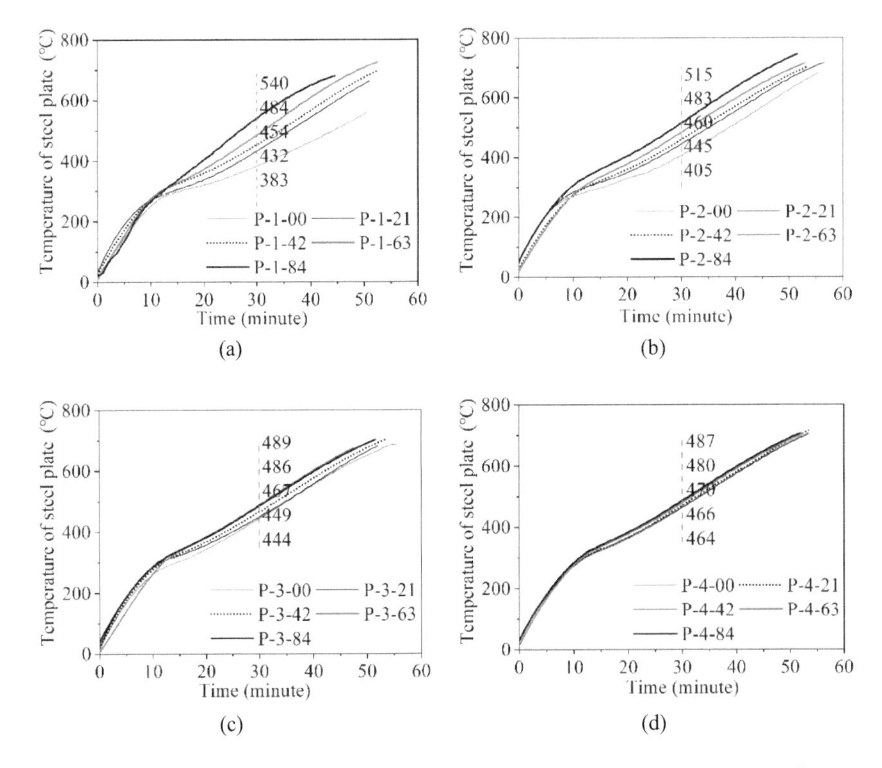

Figure 6.15 Comparison of average steel substrate temperature–time relationships for type-N specimens. (a) With one layer of topcoat. (b) With two layers of topcoat. (c) With three layers of topcoat. (d) With four layers of topcoat.

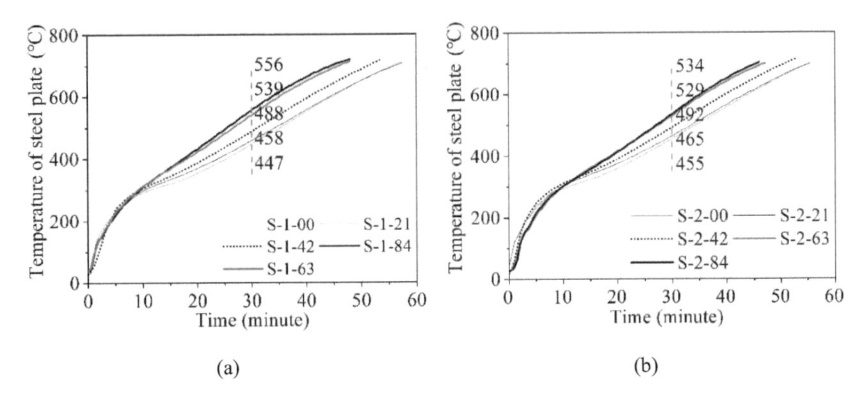

Figure 6.16 Comparison of average steel substrate temperature–time relationships for type-P specimens. (a) With one layer of topcoat. (b) With two layers of topcoat.

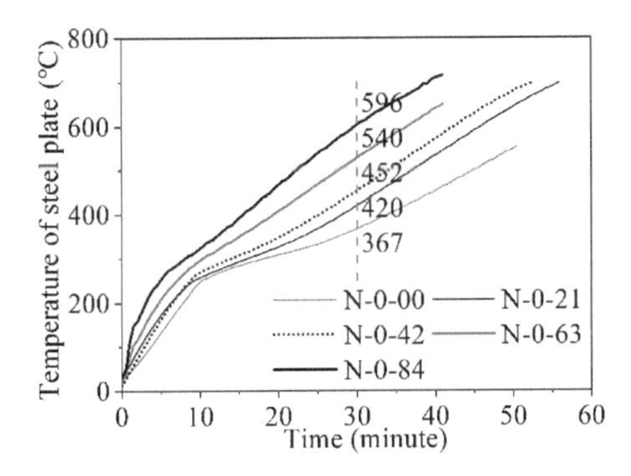

Figure 6.17 Comparison of average steel substrate temperature–time relationships for type-S specimens.

the presence of a topcoat is detrimental to intumescent coating performance due to its restraining effect in reducing coating expansion, as shown in Figure 6.14. However, after long durations of ageing, the topcoat protects the intumescent coating thereby allowing the coating to behave more effectively than one without a topcoat. For example, with four layers of a type-P topcoat, the temperature at 30 minutes of the specimen after 84 cycles of ageing was only about 24°C higher than that with no ageing, compared to a temperature increase of 229°C if there was no topcoat.

The results in Figure 6.18 were replotted in Figure 6.19 to examine the effects of the number of layers of topcoat on the temperatures of the steel substrate protected by intumescent coatings at 30 minutes of fire tests. It can be seen that using more layers of topcoat, the temperature increase of the steel substrate protected by intumescent coatings due to ageing is more reduced, compared with the case without a topcoat, because the ageing of intumescent coatings is more effectively mitigated by a topcoat.

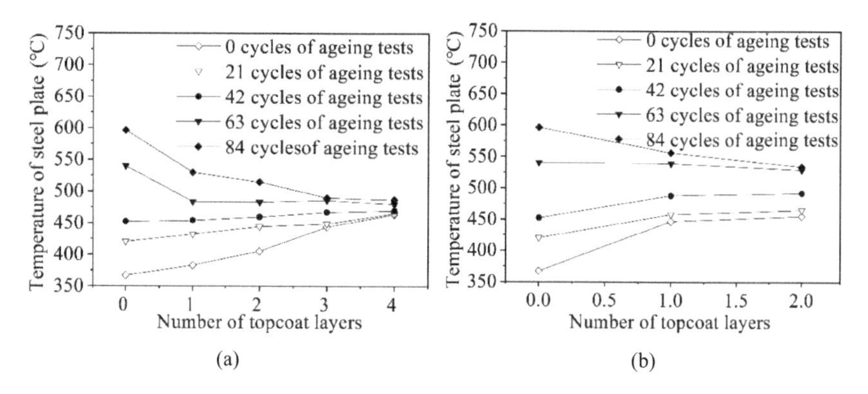

Figure 6.18 Comparison of steel substrate temperatures at 30 minutes between different ageing cycles. (a) Type-P specimens. (b) Type-S specimens.

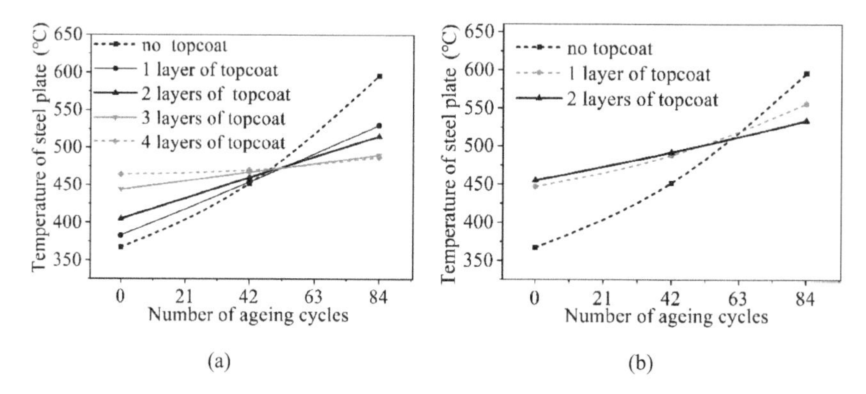

Figure 6.19 Comparison of steel substrate temperatures at 30 minutes between different numbers of topcoat layers. (a) Type-P specimens. (b) Type-S specimens.

6.3 SUMMARY

This chapter has presented comprehensive studies on the effect of topcoats on fire protection performance and the anti-ageing properties of intumescent coatings.

Experimental studies were conducted on intumescent coating specimens applied with topcoats. It was found that a topcoat may restrain the thermal expansion of intumescent coatings. However, the topcoat had little effect on the insulative properties of waterborne intumescent coatings under post-flashover fire, but there was some detrimental effect when exposed to large space fire. Nevertheless, the effect of a topcoat on the fire protection performance of solvent-based intumescent coating is notable, regardless of the fire feature. In addition, the tests also demonstrated that the effect of topcoats decreases as the intumescent coating thickness increases.

The effects of topcoats on the anti-ageing properties of intumescent coatings were also studied through experiments. It was found that topcoats can effectively protect these coatings against ageing. With topcoats, the fire-induced steel substrate temperatures with intumescent coatings after different cycles of ageing were similar to that of the specimens without a topcoat and were not increased due to the ageing of the insulative coatings. This indicates that the intumescent coatings with a topcoat would be able to maintain their fire protection function throughout their service life. Further, it is confirmed that one or two layers at most of topcoats are sufficient for guarding intumescent coatings against ageing.

REFERENCES

Chinese Engineering Construction Society Standardization. 2006. *Technical Code for Fire Safety of Steel Structure in Buildings*. CECS 200: 2006. Beijing: Chinses Engineering Construction Society Standardization.

EOTA (European Organization for Technical Assessment). 2017. *European Assessment Document-Fire Protection Products, Reactive Coatings for Fire Protection of Steel Elements*. EAD 350402-00-1106: 2017. Brussels: EOTA.

ISO (International Organization for Standardization). 1975. *Fire Resistance Tests-Elements of Building Construction*. ISO 834: 1975. Geneva: ISO.

Li, Guo-Qiang, Han, Jun, Lou, Guo-Biao, and Wang, Yong-Chang. 2016. "Predicting Intumescent Coating Protected Steel Temperature in Fire Using Constant Thermal Conductivity." *Thin-Walled Structures* 98: 177–184.

Morys, Michael, Illerhaus, Bernhard, Sturm, Heinz, and Schartel, Bernhard. 2017. "Variation of Intumescent Coatings Revealing Different Modes of Action for Good Protection Performance." *Fire Technology* 53(4): 1569–1587.

Shuklin, S. G., Kodolov, V. I., and Klimenko, E. N. 2004. "Intumescent Coatings and the Processes that Take Place in Them." *Fibre Chemistry* 36(3): 200–205.

Wang, Ling-Ling, Wang, Yong-Chang, and Li, Guo-Qiang. 2013. "Experimental Study of Hydrothermal Ageing Effects on Insulative Properties of Intumescent Coating for Steel Elements." *Fire Safety Journal* 55: 168–181.

Wang, Ling-Ling, Wang, Yong-Chang, Li, Guo-Qiang, and Zhang, Qian-Qian. 2020. "An Experimental Study on the Effects of Topcoat on Ageing and Fire Protection Properties of Intumescent Coatings for Steel Elements." *Fire Safety Journal* 111: 102931.

Wang, Zhen-Yu, Han, En-Hou, and Ke, Wei. 2006. "Effect of Nanoparticles on the Improvement in Fire-Resistant and Anti-Ageing Properties of Flame-Retardant Coating." *Surface and Coatings Technology* 200(20–21): 5706–5716.

Xu, Qing, Li, Guo-Qiang, Jiang, Jian, and Wang, Yong-Chang. 2018. "Experimental Study on the Influence of Topcoat on Insulation Performance of Intumescent Coatings for Steel Structures." *Fire Safety Journal* 101: 25–38.

Chapter 7

Predicting the temperatures of steel substrates with intumescent coatings under non-uniform fire heating conditions

With the insulative properties of intumescent coatings presented in previous chapters, the temperature elevation of protected steel substrates exposed to uniform fire heating conditions can be easily evaluated with the simple one-dimensional heat transform principle, as shown in Equation (2.1). However, if the fire heating condition is non-uniform, such as a localized fire scenario (Zhang and Li 2012, 125; Byström et al. 2014, 148; Xu et al. 2020, 6–7), the evaluation of the fire-induced temperatures of the steel substrates becomes complicated due to the three-dimensional thermal conductivity of these substrates.

This chapter presents an approach to predicting steel substrate temperatures with intumescent coatings under non-uniform fire heating conditions. The feasibility of the three-stage thermal conductivity model for intumescent coatings under localized fire is investigated and validated to be used for predicting complicated fire-induced steel substrate temperatures (Xu 2021, 71–82).

7.1 STEEL TEMPERATURE CALCULATION METHOD

The well-established lumped mass method employed in European code (CEN 2005, 40) and Chinese code (GB 2017, 26) for calculating the temperature of a protected steel member actually assumes the steel temperature induced by fire is uniform over the complete member. This lumped mass method is effective at calculating the uniform temperature of a protected steel member exposed to the ISO 834 standard fire (ISO 2015, 12) or a uniform large space fire (Han et al. 2019, 11). However, the temperature of such a member under localized fire is not uniformly distributed over the cross-section and the length of that member (Ramesh et al. 2020, 5–6). To deal with non-uniform temperature distribution along the length and through the cross-section of a steel member exposed to localized fire, the member is firstly divided into a number of segments along the length, and then each segment is further divided into a number of plate elements through the cross-section. For each individual plate element of any segment, the lumped mass method is applied (Xu et al. 2021, 3–4).

DOI: 10.1201/9781003287919-7

Equation (2.3) is based on the thermal boundary condition that assumes that the intumescent coating surface temperature is the same as the fire gas temperature, thereby ignoring the thermal resistance of the gas layer at the interface of the fire and intumescent coating. This assumption is adopted in design codes such as EN 1993-1-2 (CEN 2005, 40) and GB51249 (GB 2017, 26) because the thermal resistance of the interface can be ignored when the fire source is close to the coatings and the gas temperature is high. For intumescent coatings located far away from the fire source, the thermal resistance of the interface may have some effect on the fire-induced protected steel temperature. However, in this case, the temperature is generally low (less than 200°C) and not critical to the safety of the members in the fire.

7.2 DIVISION OF STEEL MEMBERS EXPOSED TO NON-UNIFORM FIRE HEATING CONDITIONS

7.2.1 Division of steel members into segments

The division of a steel member along its length may be done according to the gas temperature distribution along the member. In regions along the steel member where the surrounding fire gas temperature is similar, just one or a small number of segments are sufficient. Where there are high gas temperature gradients, more segments are necessary. Figure 7.1 (Xu et al. 2021, 3) gives an example of dividing a steel member into segments along its length according to the surrounding gas temperature distribution. Within each length, the temperature is assumed to be uniform, being that at the mid-point of the segment.

7.2.2 Division of a cross-section into plates

Each segment along the steel member's length can be divided into a number of plates via cross-section of the segment. The temperature distributed in each plate of any segment is assumed to be uniform, and there is no heat conduction between the plates. Figure 7.2 illustrates how to divide a steel section into individual steel plates.

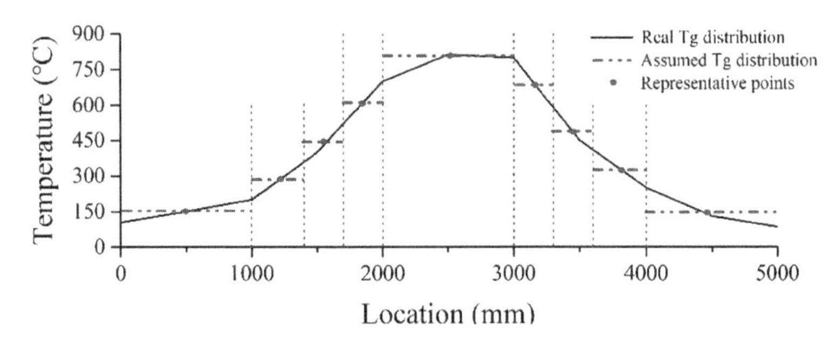

Figure 7.1 Division of a steel member into segments along its length.

For a steel plate heated from one side only (such as those of the tube member in Figure 7.2), the section factor for the plate element can be calculated as $1/t$, where t is the plate thickness. For a steel plate heated from both sides with the same fire temperature, the section factor is $2/t$. However, under localized fire exposure, it is possible for a steel plate to be exposed to different fire temperatures on the two sides, such as the flange and web plates of an I-shaped section as shown in Figure 7.2. An equivalent section factor for the steel plate can be calculated based on the principle of heat balance, that is the total heat input into the steel element from the two sides is equal to the heat for changing the steel temperature:

$$\rho_s c_s V \frac{\Delta T_s}{\Delta t} = \frac{(T_1 - T_s) A_1}{(d_p / \lambda)} + \frac{(T_2 - T_s) A_2}{(d_p / \lambda)} = \frac{(T_1 - T_s) A_e}{(d_p / \lambda)} \tag{7.1}$$

where A_1 is the surface area per unit length of side 1 of the member [$m^2\ m^{-1}$]; A_2 is the surface area per unit length of side 2 of the member [$m^2\ m^{-1}$]; A_e is the equivalent surface area for the whole steel plate when assuming that the surrounding gas temperature is T_1 [$m^2\ m^{-1}$]. For convenience, $T_1 > T_2$.

Because the steel temperature inside the plate is assumed to be uniform, the thermal conductivities of intumescent coatings on both sides are the same. Therefore, the equivalent section factor can be obtained as:

$$\frac{A_e}{V} = \frac{A_1}{V} + \frac{A_2}{V} \frac{T_2 - T_s}{T_1 - T_s} \tag{7.2}$$

Because the steel and surrounding gas temperatures change with time, Equation (7.2) indicates that the equivalent section factor is a varying value.

For a steel plate, $A_1/V = A_2/V = 1/t$, therefore Equation (7.2) is simplified to:

$$\frac{A_e}{V} = \frac{1}{t}\left(1 + \frac{T_2 - T_s}{T_1 - T_s}\right) \tag{7.3}$$

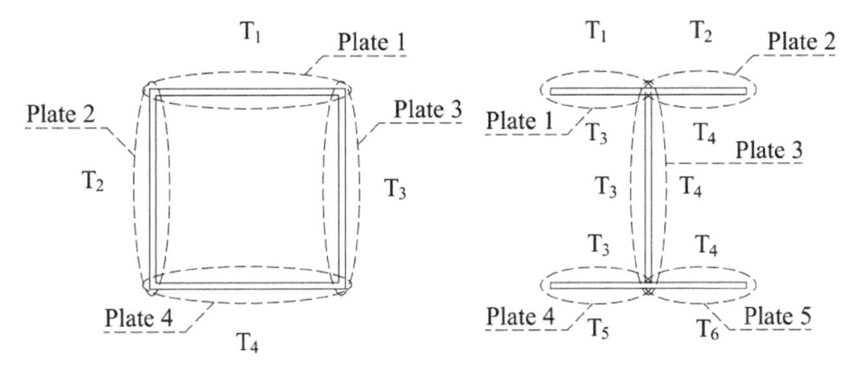

Figure 7.2 Division of a cross-section into plate elements.

It should be pointed out that when only one side of the plate is exposed to fire and the other side is not, that is $T_2 = 0$, the section factor from Equation (7.2) does not become 1/t as for the one-side exposed case. This anomaly is a result of ignoring heat dissipation on the unexposed side when using 1/t as the section factor, that is assuming the second term in the brackets of Equation (7.3) is zero. Using Equation (7.2) would give a section factor less than 1/t, which is generally conservative for estimating the protected steel temperatures exposed to a localized fire (Xu et al. 2021, 3–4).

7.3 GAS TEMPERATURE DISTRIBUTION OF LOCALIZED FIRES

A series of localized fire tests were conducted on six steel member specimens protected by intumescent coatings as presented in Chapter 4. Figure 7.3

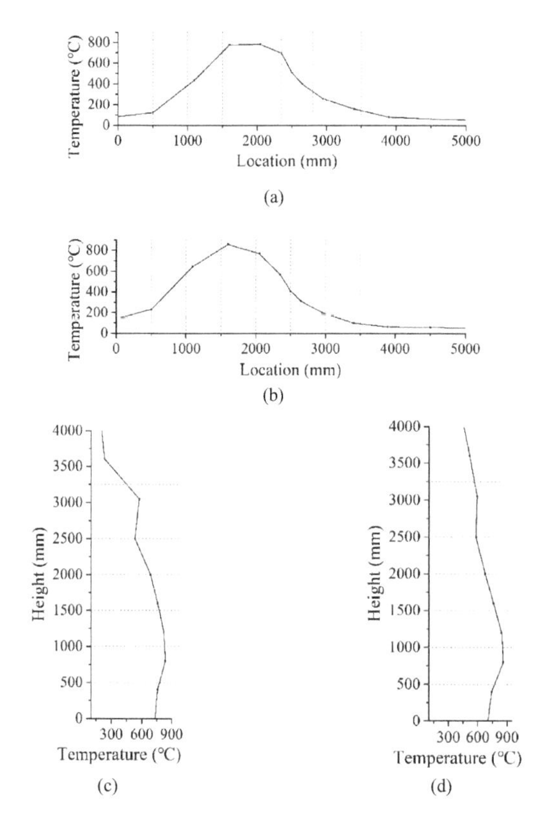

Figure 7.3 Recorded average surrounding gas temperature distributions along the length of the surface with the highest temperature for tested specimens exposed to localized fires. (a) Beam specimen with solvent-based intumescent coating. (b) Beam specimen with waterborne intumescent coating. (c) Column specimen with solvent-based intumescent coating. (d) Column specimen with waterborne intumescent coating.

(Xu et al. 2021, 7) shows the recorded gas temperature distributions close to the tested specimens, which are used for the calculation of the protected steel temperatures of the tested specimens exposed to localized fires.

7.4 TEMPERATURE DISTRIBUTIONS OF STEEL MEMBERS EXPOSED TO LOCALIZED FIRES

7.4.1 Comparison of steel temperature distributions along member length

Figure 7.4(a–d) (Xu et al. 2021, 8–9) shows comparisons of steel temperature distributions along the length of the tested specimens on the side with the highest steel temperature respectively obtained between calculation and measurement at two different times: 700 s and at the end of the test. The satisfactory agreement between the calculated estimation and test measurement confirms that the simplified temperature calculation method proposed is able to predict the steel temperature distributions along the member length, thereby confirming the negligible effects of heat conduction along the member length.

Table 7.1 (Xu et al. 2021, 12) compares the maximum steel temperatures of the tested specimens under localized fires. The maximum difference between the results obtained from the test measurement and calculation for the protected steel substrate is only 12.7%, which is very satisfactory considering the challenge of the changing insulation properties of intumescent coatings. Within the maximum steel temperature segments, the difference between the recorded steel temperature and the simplified calculation is less than 10% for most of the locations.

7.4.2 Comparison of steel temperature distributions across member section

For a selection of members' cross-sections, marked out by green circles in Figure 7.4(a–d), which represent the highest temperature region of each member where intumescent coatings are most likely to have experienced all three stages of behavior, a comparison is made for temperature distributions across the section, as shown in Figure 7.5(a–d) (Xu et al. 2021, 10).

It can be seen that the steel temperatures across the overall section obtained from a simplified calculation agrees with the recorded values from tests very well, the difference of which is less than 10% for most locations. This comparison also confirms that the simplified calculation method proposed is able to accurately predict the steel temperature distributions in the cross-section of protected steel members with intumescent coatings under localized fire.

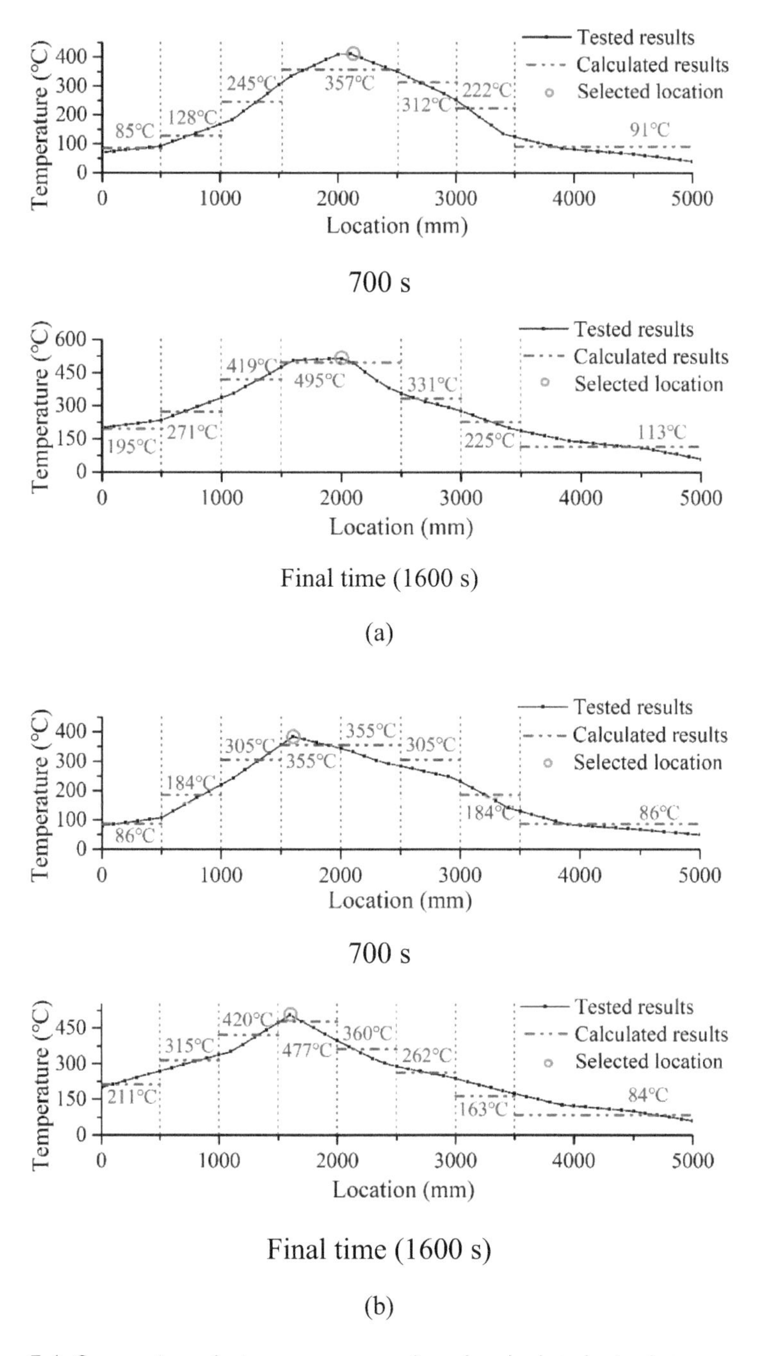

Figure 7.4 Comparison between measured and calculated steel temperature distributions along specimen length on the surface of the side with the highest temperature. (a) Beam specimen with solvent-based intumescent coating (Specimen BS). (b) Beam specimen with waterborne intumescent coating (Specimen BW).

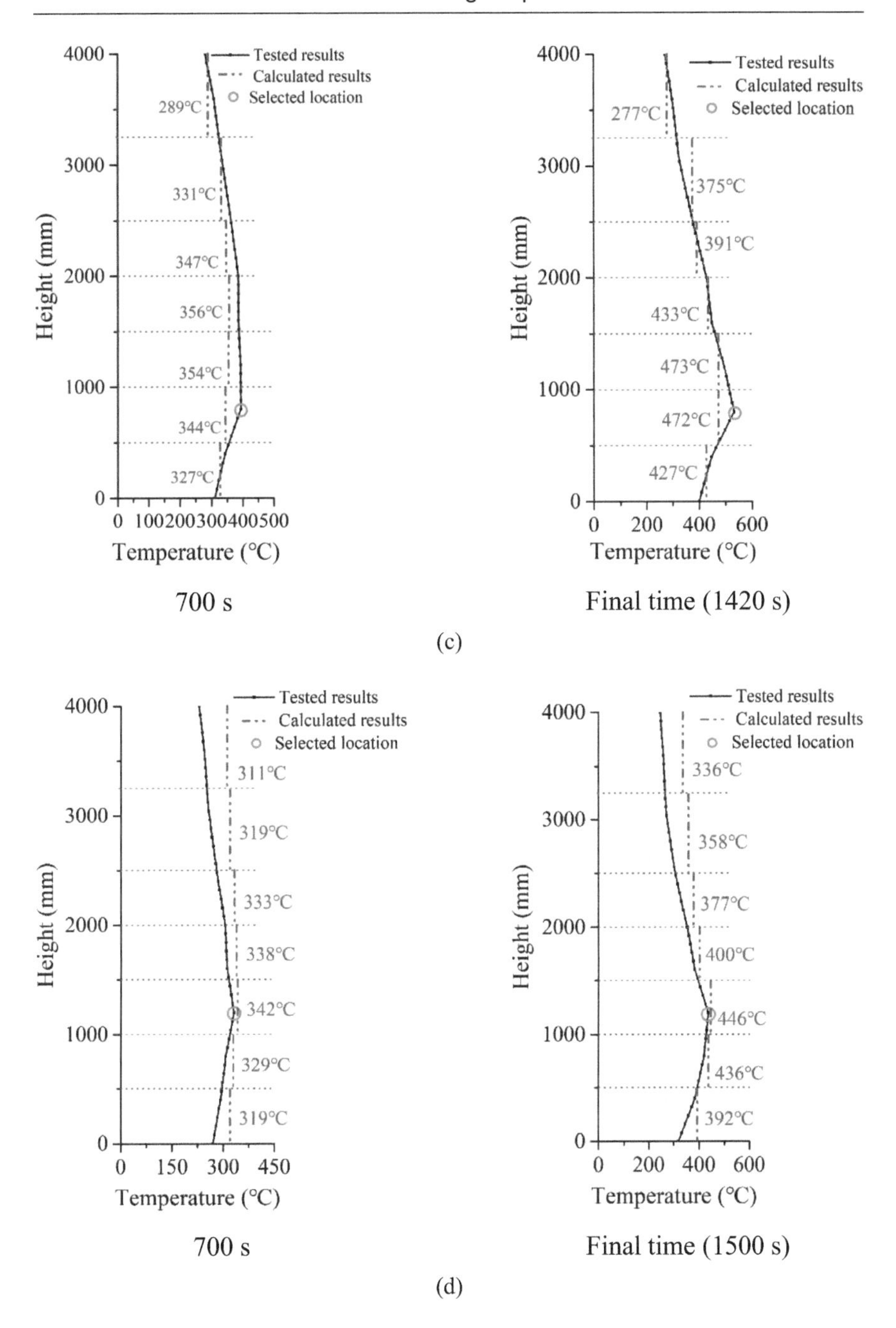

Figure 7.4 (Continued) Comparison between measured and calculated steel temperature distributions along specimen length on the surface of the side with the highest temperature. (c) Column specimen with solvent-based intumescent coating (Specimen CS). (d) Column specimen with waterborne intumescent coating (Specimen CW).

Table 7.1 Comparison of steel temperatures in the highest temperature segment

Specimen	*Segment length for simplified calculation*	*Relative location within the segment*	*Measured from tests T_s (°C)*	*Calculated T_s (°C)*	*Difference between measured and calculated (%)*
BS	1000 mm	0	476.4	495	−3.8
		0.5L	515.5		4.1
		L	356.4		−28
		Maximum temperature	515.5		4.1
		Average temperature	466.5		−5.8
BW	500 mm	0	474.5	477	−0.4
		0.5L	465		−2.4
		L	397.8		−16.5
		Maximum temperature	505.3		6.0
		Average temperature	455.38		−4.4
CW	500 mm	0	427.05	446	−4.2
		0.5L	428.3		−4.0
		L	395.0		−11.4
		Maximum temperature	435		−2.5
		Average temperature	419.8		−5.9
CS	500 mm	0	467.8	472	−0.9
		0.5L	521.1		10.4
		L	513.6		8.8
		Maximum temperature	531.8		12.7
		Average temperature	506		7.2

L: length of the corresponding segment.

7.4.3 Comparison of steel temperature–time curves

Besides the comparisons of the steel temperatures at a specific time during the localized fire tests, as presented in Sections 7.4.1 and 7.4.2, Figure 7.6(a–d) (Xu et al. 2021, 11) compares steel temperature–time curves at the maximum steel temperature positions of the specimens during tests, as indicated by the green circles in Figure 7.4(a–d) respectively. This comparison further confirms that the calculation method proposed in this book can provide satisfactory protected steel temperature results closely following the values measured from tests throughout the entire fire heating phase. In particular, after the steel temperature has exceeded 400°C, when the intumescent coating has fully expanded and steel structural behavior is critically affected by the elevated steel temperature, differences of the protected steel temperatures are obtained by calculation and test measurement decrease.

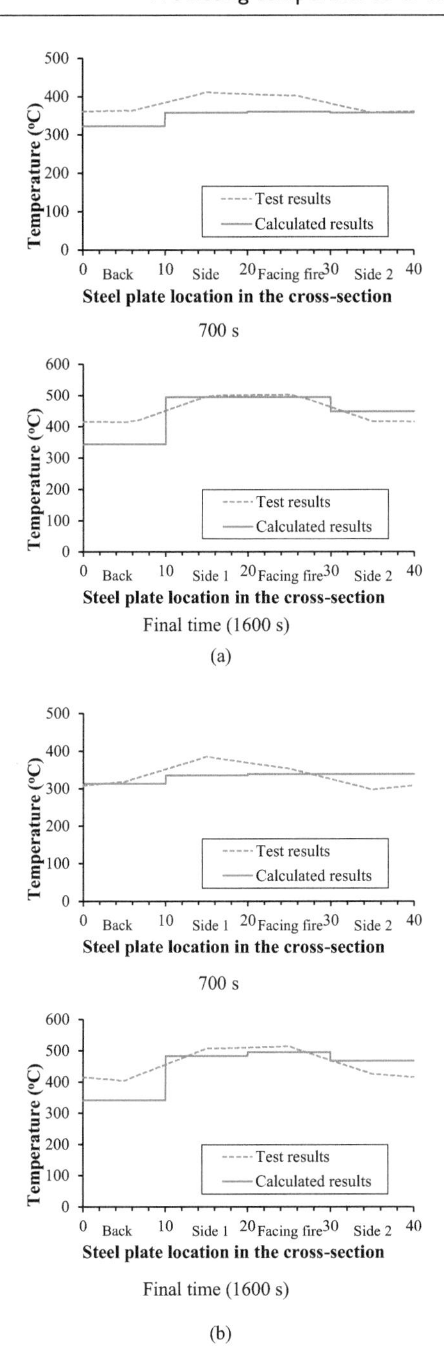

Figure 7.5 Comparison between measured and calculated steel temperature distributions in the cross-section at the highest temperature locations of specimens. (a) Beam specimen with solvent-based intumescent coating (Specimen BS). (b) Beam specimen with waterborne intumescent coating (Specimen BW).

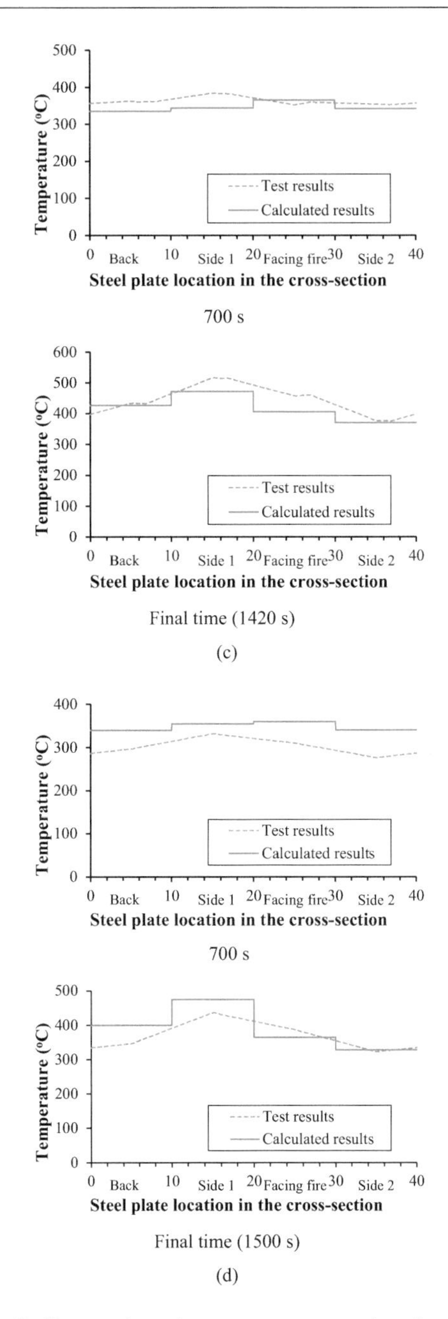

Figure 7.5 (Continued) Comparison between measured and calculated steel temperature distributions in the cross-section at the highest temperature locations of specimens. (c) Column specimen with solvent-based intumescent coating (Specimen CS). (d) Column specimen with waterborne intumescent coating (Specimen CW).

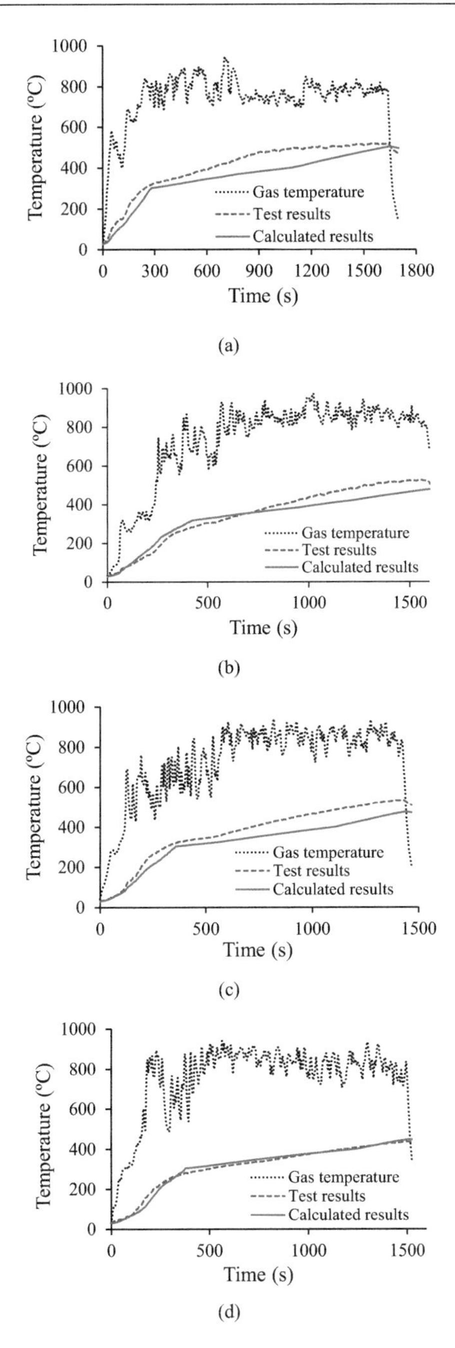

Figure 7.6 Comparison between measured and calculated steel temperature–time curves at the highest steel temperature positions. (a) Beam specimen with solvent-based intumescent coating (BS). (b) Beam specimen with waterborne intumescent coating (BW). (c) Column specimen with solvent-based intumescent coating (CS). (d) Column specimen with waterborne intumescent coating (CW).

7.5 SUMMARY

A steel member can be divided into a number of segments and further the segments can be divided into a number of plate elements. With such division, the temperature distributions of the steel members protected with intumescent coatings under localized fire can be obtained by applying the lumped mass method for calculating the fire-induced temperature of each plate element. This approach is validated with good agreement between the results of the calculation and measurement from the localized fire tests.

REFERENCES

Byström, Alexandra, Sjöström Johan, Wickström Ulf, Lange David, and Veljkovic, Milan. 2014. "Large Scale Test on a Steel Column Exposed to Localized Fire." *Journal of Structural Fire Engineering* 5(2): 147–160. https://doi.org/10.1260/2040-2317.5.2.147.

CEN. 2005. *EN 1993-1-2:2005.Eurocode 3: Design of Steel Structures — Part 1–2: General rules — Structural Fire Design*. Brussels: European Committee for Standardization.

GB. 2017. *GB51249-2017. Code for Fire Safety of Steel Structures in Buildings*. Beijing: China Planning Press.

Han, Jun, Li, Guo-Qiang, Wang, Yong C., and Xu, Qing. 2019. "An Experimental Study to Assess the Feasibility of a Three Stage Thermal Conductivity Model for Intumescent Coatings in Large Space Fires." *Fire Safety Journal* 109: 102860. https://doi.org/10.1016/j.firesaf.2019.102860.

ISO. 1999. *ISO 834–1: 1999, Fire-Resistance Tests — Elements of Building Construction —Part 1: General Requirements*. Switzerland: ISO.

Ramesh, Selvarajah, Choe, Lisa, and Zhang, Chao. 2020. "Experimental Investigation of Structural Steel Beams Subjected to Localized Fire." *Engineering Structures* 218: 110844. https://doi.org/10.1016/j.engstruct.2020.110844.

Xu, Qing. 2021. *Study on the Performance of Steel Structures Protected by Intumescent Coatings under Localized Fires*. Shanghai: Tongji University.

Xu, Qing, Li, Guo-Qiang, and Wang, Yong C.. 2021. "A Simplified Method for Calculating Non-uniform Temperature Distributions in Thin-Walled Steel Members Protected by Intumescent Coatings under Localized Fires." *Thin-Walled Structures* 162: 107580. https://doi.org/10.1016/j.tws.2021.107580.

Xu, Qing, Li, Guo-Qiang, Wang, Yong C., and Bisby, Luke. 2020. "An Experimental Study of the Behavior of Intumescent Coatings under Localized Fires." *Fire Safety Journal* 115: 103003. https://doi.org/10.1016/j.firesaf.2020.103003.

Zhang, Chao, and Li, Guo-Qiang. 2012. "Fire Dynamic Simulation on Thermal Actions in Localized Fires in Large Enclosure." *Advanced Steel Construction* 8(2): 124–136.

Index

Pages in *italics* refer figures and **bold** refer tables.